Julius George Medley

**India and Indian Engineering**

Three Lectures Delivered at the Royal Engineer Institute, Chatham, in July,

1872

Julius George Medley

**India and Indian Engineering**
*Three Lectures Delivered at the Royal Engineer Institute, Chatham, in July, 1872*

ISBN/EAN: 9783337038892

Printed in Europe, USA, Canada, Australia, Japan

Cover: Foto ©berggeist007 / pixelio.de

More available books at **www.hansebooks.com**

# INDIA AND INDIAN ENGINEERING.

## THREE LECTURES

DELIVERED

AT THE ROYAL ENGINEER INSTITUTE, CHATHAM,

In JULY, 1872.

BY

JULIUS GEORGE MEDLEY,

LIEUT.-COLONEL, ROYAL ENGINEERS; ASSOC. INST. C.E.;
FELLOW OF THE CALCUTTA UNIVERSITY;
PRINCIPAL, THOMASON CIVIL ENGINEERING COLLEGE, ROORKEE.

LONDON
E. & F. N. SPON, 48, CHARING CROSS.
NEW YORK:
446, BROOME STREET.
1873.

LONDON: PRINTED BY W. CLOWES AND SONS, STAMFORD STREET AND CHARING CROSS.

# SYLLABUS OF LECTURES.

INDIA — its area — physical features — climate — scenery. THE PEOPLE — Bengalees — languages — Hindooism — caste — conservatism of the East — the Mahomedans—Sikhs—Parsees. THE ENGLISH IN INDIA — their difficulties — the Anglo-Indian career — the mutiny—Christianity in India—arts and manufactures —general character of the people. ANGLO-INDIAN LIFE—in the station—in tents—cost of living—society in India—travelling—a tour in India.

The Indian Government—the Public Works Department. ROORKEE—the Thomason College—the Sappers and Miners — the workshops — career of a Royal Engineer—military engineering—miscellaneous duties of the Indian engineer—financial aspects of the Public Works Department—overseers—native subordinates—workmen.

BUILDING MATERIALS—stone—bricks—tiles—limes—timber—iron—wages and rates—weights and measures—absence of plant—water-raising machines—carts. FOUNDATIONS—well-cylinders—Indian rivers.

BARRACKS—difficulties of ventilating and cooling—private houses—churches—other buildings. BRIDGES—temporary—permanent—waterway. ROADS—metalling—hill roads. RAILWAYS—various lines—the permanent way—traffic arrangements. IRRIGATION WORKS—their importance—the Ganges Canal—crops and soils—design of canals—the head—velocity of stream—falls and rapids—drainage works—irrigation details—Madras weirs—tanks. RIVER WORKS—inundations—spurs.

INDIAN SURVEY DEPARTMENT — the Great Trigonometrical survey — Topographical survey — Revenue survey.

# INDIA AND INDIAN ENGINEERING.

In proposing to deliver the short series of Lectures which I commence this evening, I had two objects in view; First, to interest you in the work which your brother officers are doing at the other end of the world, and which I think is little understood or appreciated in this country; Secondly, to give to those amongst you who are likely to proceed to India some useful information about the country itself, the nature of the work you will be called upon to undertake, and the special subjects of study to which it is desirable on that account to direct your attention.

In the present lecture, I shall endeavour to give you some idea of the physical features of India, its climate, its people, and of the peculiarities of Anglo-Indian life. In the other lectures, I shall say something of the Government, and the great Department of State by which public works are executed, and of the special duties and probable career of the Royal Engineer officers who are there employed; and shall then pass on to the materials and modes of construction with which the engineer is called upon to deal, and those specialities which distinguish his work from English practice.

India, then, is about as large as Europe without

Russia. A line drawn from Cape Comorin at the south to Peshawur in the north will measure about 2000 miles; another line drawn from Kurrachee on the west coast to Calcutta will measure some 1500 miles; the total area of the whole peninsula, including British Burmah, is about 1,500,000 square miles, of which 900,000 are directly under British rule, while the remainder, though nominally under native governments, is more or less subject to us.

This vast area of country comprises almost every variety of physical configuration—lofty mountains and low hills; well-cultivated, alluvial plains, arid deserts, great forests, marshy swamps and dense jungles; long, broad rivers, numerous hill torrents, wide and deep nullahs. The varieties of climate to be found in this great continent are also numerous; for while the plains of Upper India are for several months parched up with a fiery heat, the summits of the Himalayahs are covered with perpetual snow; and while the rainfall of Sindh seldom exceeds four or five inches annually, there is a place in Assam called Cherra Poonjee, well known to geographers as the rainiest place on the earth's surface, the annual fall amounting to 650 inches.

The popular idea of India is that it is an extremely hot country, and speaking generally, the popular idea is correct. But the nights in Northern India are often excessively cold, and I have many times seen ice half an inch thick on the roadside puddles in the Punjab, while the hill stations at Simla, Mussorie, and elsewhere, 7000 feet above the sea level, are covered with snow in January and February.

In Upper India, the part with which I am best acquainted, we enjoy a climate which for four months in the year—November, December, January, and February—is probably unequalled in the world both for health and pleasure: bright skies, a sun hot indeed, yet not too hot for exercise all day long, and nights cold, dry, and bracing, with a clear, still atmosphere, make an almost perfect climate. In October, March, and April, the mornings, evenings, and nights are still delightful, though the heat out of doors in the daytime is great. For the remaining five months, the climate, to a European at least, is simply detestable. You have either fierce hot winds like the blast from a glass furnace, with clouds of dust; or else a moist, stagnant atmosphere like that of a continuous vapour bath, and excessively depressing. The nights are rather worse than the days, and life is only bearable inside large and lofty rooms and under swinging punkahs. In Southern India, there is less extreme heat, but more moisture, and no real cold weather.

Yet the climate, with proper precautions and temperate habits, is by no means unfavourable to the European constitution, except in peculiar cases. As a rule, men now return from India looking much the same as their English contemporaries, and those whose minds are well employed and whose bodies get a fair share of exercise, are as healthy as their fellow-countrymen whose lot is cast in England or the

Colonies. Out of eight Engineer officers who left Chatham with me twenty-three years ago to go to India, six are now alive, and five out of the six are strong, healthy men. Nor is this at all an exceptional case; indeed, when an Anglo-Indian reaches a certain age he seems to live for ever, though the popular idea that this is because the Indian sun has dried him up into a mummy, is not founded on fact.

If I were asked whether India was a very beautiful country, I should reply that in general it is not, but that it has some of the finest scenery in the world. In travelling up the main line of railway, for instance, from Calcutta to Peshawur, your road lies for 1000 miles of that distance over a country that is one dead level, without even a hillock to break the monotony. If your journey is made (say) in March, as far as the eye can reach it rests on an enormous sea of wheat, diversified by groves of mangoe trees, and mud villages, or brick-built towns. No crystal streams,—no clear lakes,—no undulating downs,—no parks or country houses,—not even a grass field. Yet the rich cultivation, and the general signs of prosperity amongst the dense population are at least pleasing to the philanthropist; and if we leave the railway at Umballa and travel for forty miles eastward, we find ourselves amongst the dark pine forests, the mountain torrents, and the craggy heights of the Himalayahs, while their gigantic tops covered with eternal snow, 10,000 feet higher than "the monarch of mountains," look down upon us in their calm and solemn grandeur.

Nor are the great forests and mountain ranges of

Central India without much beauty; while the magnificent harbour of Bombay and other sea views on the coast show that India is not wanting in many of the charms of marine landscape.

But it is time that I should speak of the people of the country. Those who have never been in India often form their ideas, (very naturally), from the few natives of India they have met in this country; but these are a small and very peculiar class, and are by no means fair specimens, not merely of the whole population, but even of their own province.

Let me, however, at once give you a few figures, which will show how unsafe it is to generalize from a few instances. India is inhabited by about 200 millions of people, speaking at least eleven totally distinct languages, and innumerable dialects, and differing amongst each other in features, character, and social customs, quite as much as the Russian or Spaniard does from the Englishman.

As, therefore, I have told you not to form a judgment of the whole from a few isolated and exceptional instances, I shall avoid falling into the same error, and only talk about the races with which I am personally acquainted.

The Bengalees, *i. e.* the inhabitants of Bengal proper, have been those who have benefited most intellectually by their contact with the English. They are quick-witted and clever; many are excellent English scholars, and make admirable clerks and accountants. Many have risen to high positions in the public service: one sits on the Bench as a Judge of the High Court; several have come to England and fairly won their place in the Indian Civil Service, in competition with Englishmen. But no Bengalee serves in the army, or enters into any pursuit or amusement from choice which requires bodily activity and strength: his physical organization is of the feeblest; for centuries he has been ruled by the stronger races from the North, and it is to be feared that his moral organization but too often follows the law of the physical.

As we go north, after leaving Bengal, we find ourselves amongst a more manly race—the stalwart Jat, the manly Rajpoot, the warlike Sikh, and the fierce and treacherous Puthan. These are the men with whom we wrestled for the empire of the East, and who now recruit our best native regiments,—who helped to plant the British flag on the towers of Pekin and the heights of Magdala.

Nor are these men at all deficient in intellectual power, though they have been less quick than the Bengalee to appreciate the advantages of education. But the Northern colleges and schools are now crowded with students, and even the frontier chieftains, who once thought it disgraceful that a son of theirs should wield a pen instead of a sword, have given in their adherence to the new-fangled ways of their conquerors.

The chief languages spoken in India are Tamil, Teloogoo and Canarese in the south; Mahratta and Guzerati in the west; Bengali, a dialect of Hindi, in

Bengal; Oordoo or Hindustani in the North-west Provinces; Punjabee, a dialect in the Punjab; Burmese in Burmah; and Pooshtoo, the language of the Afghans. Sanscrit, as you probably know, is a dead language, in which the Shasters, or Hindoo scriptures, are written. Arabic is that of the Koran, read in India, but not spoken. Persian is only used by the best-educated people at the native courts. The Hindustani, or Oordoo, is the *lingua franca* current to a certain extent over the empire, at least, amongst the people with whom we chiefly come in contact, but scarcely understood by the people generally out of Hindustan, *i.e.* the Northwest Provinces. It is a mixture (not a compound,) of words from the Hindi, Persian, Arabic, and Sanscrit, not unlike French in many of its characteristics, and easy to speak, but difficult to read and write. It is written both in the Persian and Hindi characters.

It is absolutely necessary for every Anglo-Indian to acquire a certain colloquial proficiency in it; for English-speaking natives are rare, and when found in the ranks of domestic servants, do not bear the best of characters.

The people of Upper India generally are a goodlooking race, with well-formed features and good figures. The complexion varies very much, but is generally brown or dark olive—very rarely black. Many of the men are strikingly handsome, but the best-looking women are generally secluded at home, and it is only occasionally that one sees a real Eastern beauty. I need hardly tell so intelligent an audience that both the Indian and English races belong to the same great Aryan stock, and that the people of whom I am speaking have nothing in common with the low-developed, barbarous races of Southern or Central Africa, or Australasia. On the contrary, they have a language, a religion, a code of laws, and a civilization, considerably older than our own, and which, though now degraded from their original purity, yet exist in full force amongst the great bulk of the people, and by their wonderful conservatism and adaptation to the requirements of Eastern life, bid fair to maintain their ascendancy for many generations to come, and to set European innovations at defiance.

As you know, of course, the vast majority of the people of India are Hindoos by religion. Boodhism, once prevalent in India as it is now in China, has disappeared, and Brahminism prevails. I have no time to enter into any learned dissertation on the Hindoo tenets, but I must allude to one of its most distinguishing features—that of Caste—because it has more practical bearing on the every-day life of the people than all the rest of the tenets put together, and because it is also generally misunderstood.

Everyone probably knows that the original division of mankind, according to the Vedas, was into the four castes of the Brahmins, or priests; the Rajpoots, or warriors; the Ksatryahs, or writers; and the Sudras, or low-caste men. But this division has been so modified and altered, that it has practically disappeared. Brahmins are soldiers, traders, or cultivators, as well as priests; Rajpoots are cultivators rather than warriors, while all four castes have been divided and subdivided into innumerable petty castes, which, as a rule, are identical with the trade or calling of their votaries. Thus, a man who is a carpenter, will bring up all his sons to be carpenters, and so on *ad infinitum;* though this is being slowly altered where education opens out a prospect of more profitable employment in another line.

To lose caste, or be put out of caste, is as great a misfortune as ever, but there are very few offences for which a man cannot get back his caste by the payment of a few rupees, which are expended in eating and drinking by his fellow caste-men. The offences which involve loss of caste are offences against custom rather than religion, and indeed there are no people so grossly ignorant of technical religion and their sacred books as the Hindoos. The ordinary Brahmins are no better than the common people, and caste is a thing of custom and not of religious doctrine.

But if it be thought that, on that account, its hold on the people is small, the thinker has very little acquaintance with the power of custom in the East. The most bigoted Tory in England—if such a phenomenon now exists—is a Red Republican in presence of the Conservatism of the East. There you may see the land cultivated and the fields watered now as they were 2000 years ago, when the Macedonian phalanx defeated Porus on the banks of the Jhelum; there you may see "two women grinding at the mill" the corn for the daily meal, and can understand the force of the prophecy that one shall be taken and the other left. The ploughs and carts in every-day use are the same as those shown on the sculptures of Egypt or Assyria. The unleavened cakes that Sara prepared on the hearth for the angels were exactly similar to those your Indian servants now give you if you want a hasty meal. You see hundreds of men every morning sleeping outside their houses, and "taking up their beds and walking," by the simple process of rolling up their light cotton mattress under one arm, or carrying it and their light bamboo bedstead on their heads together. The women draw water from the well, and poise the same shaped vessels on their heads that Rebekah did when Abraham's servant greeted her, and, but a few steps off you will see the camels kneeling down, and the men unloading their burdens.

It is this wonderful conservatism that perhaps strikes the observant traveller more than anything else in the East; which opens his eyes to a state of society utterly foreign to all his Western experience, and makes him pause to think whether he is right after all in his ideas of the advantages of civilization. Is the man of the West any happier for his railways, electric telegraphs, steam factories, and Parliamentary Governments? Here he finds people who are not in the least anxious to govern themselves; who think fifteen or twenty miles a rather long day's journey, and very seldom take that; who are content to follow their fathers' calling as a matter of course, and who shrink with horror from that restless, bustling, feverish, active life which has become a second nature to the Englishman. The fact is, that each follows out, so to speak, the law of his being, and neither has a right to dictate to the other as to how he shall find his happiness.

Before leaving the subject of the Hindoo religion, I should perhaps mention that the hideous customs of Suttee (*i.e.* burning the living wife with the dead husband), and the suicide of pilgrims under the car of Juggernath, are now things of the past, and indeed they were never sanctioned by the Hindoo sacred books. The crime of female infanticide amongst certain high-caste tribes still remains, but it too has nothing to do with the Hindoo religion; it arises from the social custom that a man of high rank is disgraced if his daughter is unmarried, while the same tyrant custom has imposed on him the necessity of spending large sums at his daughters' weddings. India is not the only country, nor are the Hindoos the only people, whom the tyranny of custom compels to extravagance and disregard of the obligations of common sense and right feeling, and even religious or moral duties.

Eight hundred and fifty years ago the Mahomedan armies overran India, and, after a series of fierce struggles, founded the empire of the Great Mogul at Delhi. Cities were sacked, their people massacred, temples and idols overthrown, and thousands of Hindoos were forcibly converted to Mahomedanism. But the fierce torrent soon spent its fury against the stolid wall of Hindooism. Few converts were made after the first few years, the new religion was even modified by the old, and at the present day the Mahomedans of India are scarcely one-tenth of the whole population, though doubtless a very important section of it. Generally speaking, they are a more manly race, as if they still possessed something of their former prestige. But Indian Mahomedanism is but a poor affair after all; it has taken from Hindooism the idea of caste, and the Turk, or even the Persian, would scarcely acknowledge his fellow-worshipper of India.

I must spare a moment to mention the Sikhs, a name well known in England five-and-twenty years ago, in connection with the fierce battles on the Sutlej, and afterwards at Chillianwala and Goojerat. At first only an insignificant sect, they were raised by persecution to importance as a faction, their founder being one of those earnest men who appear from time to time in every age, and, disgusted with the corruptions of religion, strive to erect a creed of Theism and morality in its stead. Now that their empire has been destroyed, their numbers are dwindling daily.

The Parsees of India are almost strangers and foreigners in the land like ourselves, descendants of the old Magi or Fire-worshippers of Persia, who, being driven out of their own country by persecution, have settled in India, where they form a small but very intelligent and respectable section of the community, possessing in a large degree those two rare qualities for an Eastern,—enterprise and public spirit. Several of them are, I believe, settled in London, and may be recognized by the peculiar high glazed hat which they always wear.

And in the midst of the 200 millions of dark-skinned people of the land dwell some 130 thousand British white faces, among them, but not of them, and indeed separated from them, not so much by the barriers of language, religion, and social customs, as by the far greater barrier of race, which, let philanthropists say what they will, has been created not without wise and useful purposes. The English in India are often judged harshly and unjustly on this head by their countrymen at home. We are told that we should mix more with the natives, and admit them on a footing of equality with ourselves. But how can you mix socially with men who will neither eat nor drink with you, and who would sooner see their wives and daughters dead than walking about with uncovered faces amongst strange men? The only real intimacy there can be between two races separated so far apart must be confined to those official or business relations in which there is a feeling of common interest; all beyond that must be forced and unnatural. If there is that equality between the races which some pretend, how is it we are there as rulers? I think, for my own part, that nothing is so apt to retard the advance of the weaker race, or to lessen the points of contact between the two, as the attempt to produce a forced and unnatural union, or to preach up an equality which every white man who has lived amongst the dark races knows in his heart does not exist.

I have often been asked whether the natives of India like us, and are attached to our rule, and it is rather a difficult question to answer. There cannot be much love or strong liking between people separated so completely in thought and feeling and almost every idea as we are, but there is often a very strong attachment to the individual Englishman placed in authority over, and living much amongst, them, such as an officer commanding a regiment, or a magistrate in charge of a district, an attachment which has often been severely tried (as during the days of the Mutiny), and which has often stood the trial successfully. As to liking for our Government, in the first place the great mass of the people in all probability never give the subject a thought; the present generation have no means of forming a comparison between British and native rule, and look upon the protection they enjoy for life and property as a matter of course. The better educated amongst them are more apt to resent their exclusion from the highest posts, than to be grateful for not having their throats cut and their houses plundered, as they might have been a hundred years ago.

The great drawback of our rule is, undoubtedly, that it is one of race over race. The Englishman, bred a free man, is forced into the position of a despot, and, in his endeavours to elevate his Indian subjects to the dignity of free men and the privileges of British citizens, he is rather apt to overdo the matter, and to forget the inherent differences of race; or, to put it in another form, to overlook the fact that our free ideas are the growth of several hundred years, just as their ideas are the growth of as many centuries of an entirely different history. Hence we are apt to see in India many ludicrous travesties of our public meetings, municipal institutions, and the like, at which the native attends to please his English superior, and does pretty well what he is told to do. When the native does attain to a high post, as a rule he is hated by his countrymen, who never thoroughly trust him, and would far sooner see an Englishman in his place.

Our attempts then at improvement are up-hill work, and made under all the disadvantages of the stiff and awkward part of our national character, which keeps us isolated even on the European continent. But we do our duty in India, I may fairly say, honestly and thoroughly, and if we have not gained the love, we have, at least, won the respect and confidence of our native subjects. If the motives of the Government are occasionally suspected or misrepresented, the individual Englishman at any rate is implicitly trusted. That is the real strength of our position, and if our Indian empire is ever destroyed by force, it will be through the decay in character of the individual Englishman.

It is the fashion in these days to sneer at what is called the selfish and exclusive policy of the old East India Company, and certainly no one could venture to propose now-a-days that the Indian Government should be vested with the power of granting or refusing a licence to any European who wished to visit India. Yet it is impossible not to respect the motive which caused the Company to ask for such a power in days gone by; their feeling that the stability of our rule depended, not upon brute force, but on prestige, on the belief the natives had of the superiority of our national character; and that every individual who, by loose or dishonest conduct, lowered that prestige in the eyes of the natives, was a dangerous enemy to the State, and ought to be removed. And when one sees, as unfortunately we often do see now-a-days in India, a disreputable and ragged fellow-countryman begging from house to house, or staggering about drunk in the native bazar, while the natives look upon him with mingled fear, contempt, and dislike, it is impossible not to wish that the power of deporting such wretched loafers from a country where they do incalculable harm, cannot be freely exercised.

And, looking on the reverse of the medal, we may say that what constitutes the great charm of an Anglo-Indian career is the feeling of individual responsibility and importance. The English are so few in number that everyone, whether civilian, soldier, merchant, or what not, as a rule enjoys a far higher position, and has more responsibility on his shoulders than he would have in a corresponding position at home. This applies more especially to the Government officials, charged with the administration of the country, and for that reason, scarcely any career offers such attractions as the Indian Civil Service. A young Englishman, very little past thirty, who, had he remained in England, might have thought himself lucky to be receiving his first brief, or might have been canvassing for the medical charge of a parish dispensary, finds himself governor of a district as large as three English counties, with a population of 300,000 souls, to whom he is the embodiment of the Government, and who look up to him for advice or direction on all possible and impossible subjects. He has extensive civil and criminal jurisdiction in his own court; he looks generally after all the schools in the district, superintends all the roads, bridges, buildings, jails, and municipal works; reports regularly to Government on the agriculture, trade, manufactures, statistics of every kind, and has usually some hobby of his own in addition; either starting agricultural shows, or improving the breed of horses, or lighting the towns with kerozine, or trying some new piece of machinery in the jail, which is the model manufactory of the district. His life may be lonely, for there are often not a dozen of his countrymen in the district. After his day's work is done, there is neither theatre, opera, nor concert to go to, nothing, indeed, in the shape of public amusement; but he is, in truth, too tired to care for it, and the interest and responsibility of his work compensate him for all.

He climbs gradually to a still higher position; perhaps becomes lieutenant-governor of a province as large as Germany, with 40 millions of inhabitants; retires at last from the service to spend the rest of his life in his native land, and finds his name and his services utterly unknown, not merely to the English public, but often to the very Government of which he had thought himself all this time the trusted servant.

It is this ignorance, and neglect of all services that are not done at home, except in very rare cases, which are the cause of so much irritation both in India and the colonies; a fourth-rate politician in England, a man who has no power or virtue whatever, save the vulgar "gift of the gab," is better known and more thought of than the ablest servants of the State 5000 miles away. I have often thought how much good might be done by those in power, if instead of philosophising so much on the exact relations between England and her colonies, and demonstrating so clearly that as they pay no taxes to the Imperial exchequer, they have no right to be defended by the Imperial armies, care was taken to show that the colonies were really considered to be an integral part of the empire, and that good service done there was to be rewarded as if done at home; if a few more royal visits were paid, and a few more ribbons and stars occasionally bestowed; and if some means were found, either by life peerages or otherwise, for the State to avail itself of the experience of those who had grown grey in its service in distant lands, and for those men to feel that their talents and knowledge were valued at home in the great council of the nation. And if you say that you English at home are not interested in this matter, I beg leave to differ, and ask you fairly if every time you look at a map of the world, and see the red colour all over the globe which marks the extent of the British empire, and the great dependencies which have been conquered and colonized by that little island in the north-west corner of Europe, you do not feel a glow of honest pride in the thought that you too are a citizen of that empire on which the sun never sets, and whether that feeling is to be valued in pounds, shillings and pence, or rather, whether you are not willing to pay many pounds in exchange for the right to that feeling.

It is that abominable material philosophy of a certain school of the present day, which recognises nothing as really valuable that does not touch the grosser part of our nature, which sneers at patriotism and sentiment of any kind, and makes a god of selfishness, that sometimes frightens those who watch the enormous increase of our national wealth, and the decrease of regard for national duty, and makes them tremble lest we should one day have a rude awakening to the fact that a selfish and exclusive policy is as bad for nations as for individuals.

To return to our subject from this digression.

To those who cannot look upon their Indian life from the standpoint I have mentioned, a career in India is, it must be owned, but a dreary exile; the time for making fortunes, at any rate in the Government service, is gone; those who retire have seldom much beyond their very moderate pensions, and while the cost of living has steadily increased for many years past, salaries have remained stationary, or even diminished, and work has very much increased.

c 2

Thus, there can be no doubt that an Indian career has fewer attractions than formerly, and this has been the case ever since the Mutiny—that great landmark in Indian history whose significance is not even yet recognised. That great struggle, remember, was in no sense an uprising of the people against our rule, for, if it had been, we could not have held India for an hour. But though it was, primarily, a military revolt, caused and aggravated by overweening confidence and bad management, it was secondarily, a struggle of conservatism against the further progress of western innovations; it was a protest by caste and tradition against railways, telegraphs, and national education. Attacked under every possible disadvantage, outnumbered in every direction, with our arsenals in the enemy's hands, and having to fight at the worst season of the year, that handful of the great Anglo-Saxon race turned fiercely to bay, supplied every deficiency by dauntless courage, wise policy, and heroic endurance, and broke the neck of the rebellion under the walls of Delhi, and in the residency of Lucknow, before a single fresh soldier had arrived from England.

Since the suppression of the Mutiny, our hold on the empire has been firmer than ever, but it owes less to prestige and more to actual strength. We have been less careful of respecting native opinion than before, more resolute to push on improvements, and the progress made in the last fifteen years in the material development of the country has been undoubtedly greater than in the previous fifty years. But much of the kindly feeling between the conquerors and conquered has gone, and will not soon be restored; the traditions and organizations of the Government services were destroyed, and have not yet been resettled, and there is no longer that attachment to the country that was seen in the days of old John Company, kindest and best of masters. The remedy for this is not, I think, a return to the old state of things, which is indeed impossible, but more close and intimate relations between India and England, until our native subjects feel that they are really regarded as part of the British empire. The more they visit England, and the more we visit India, the more will each understand and appreciate the other. We have no enemy now in India, except popular ignorance, and that we are doing our best to remove by the most complete system of State education that has yet been devised in any country.

And now I must touch on a subject on which many of you will, perhaps, expect some information. How about the progress of Christianity in India? Well, I fear it must be owned that it is extremely slow. I dare say I might be contradicted by many missionaries, but then I am not a missionary. I have the highest respect for them as a body; many I have known personally, and know to be able men; but undoubtedly their success, if judged by the number of converts, is very small, in Upper India at least; and though doubtless they do much good by keeping up the schools that are attached to every mission, that good has very little to do with the progress of Christianity. As translators of the Bible into the various Indian languages, they have been more successful, and many of them are amongst the most accomplished linguists of the East.

I am inclined to think that much of this ill-success is owing to the forgetfulness of how universal and comprehensive Christianity is. The best proof of that is that, having originated in the East, it has yet so completely conquered the West. But in that conquest it has in some respects assumed a Western garb, and I fear our missionaries often forget that this Western garb is not essential, and that so long as the life and doctrine of the Great Master are followed and understood, the peculiar form to be taken by the latter is a matter of little importance. This is not the place to enter fully into a discussion of this sort, though it was impossible for me to avoid it altogether. But I believe I am only echoing the opinions of many thoughtful and earnest Christians, like the late excellent Bishop Cotton, of Calcutta, in saying that our efforts should be rather directed to create a native Indian Church, than to reproduce the Church of England in India; and that controversial epistles addressed to Western Churches, and dealing with questions arising out of the doctrines of Western philosophy, are puzzling rather than edifying to a convert in India.

Of the state of the Arts and Manufactures in India, all of you can form some judgment yourselves by an inspection of the beautiful specimens collected in the Indian annexe of the International Exhibition.

Some of the once famous Indian manufactures have almost disappeared in modern times, such as the Dacca muslin, of which it was said that a full-sized dress piece could be drawn through a finger ring. Native architecture too of the present day is tawdry and meretricious. But Cashmere is still famous for its wonderful shawls, in which we know not which to admire most, the beauty of the fabric, or the exquisite patterns and harmonious contrast of colours; Agra still executes that beautiful inlaid stone-work, which is yet only one of the wonders of the Taj Mehal; Delhi and Benares send gorgeous embroideries, heavy with gold and rich in colouring; Cuttack furnishes its exquisite silver filagree work; Sealkote, its steel inlaid with gold in arabesque patterns; Bombay, its massive and curiously-carved ebony furniture. But Art can never attain to its highest development in the absence of a healthy national life, and it is to former ages we must turn for structures like some of the Hindoo Temples, or the great mosque at Delhi, or "the Dream in Marble" at Agra (the Taj Mehal), and even the artistic manufactures I have named are legacies from the past, that are apt to degenerate at the present day into a grotesque copying of European designs.

Yet there is an indwelling spirit of artistic grace in the East that will not easily die, which you see in the instinctive choice of colours in the clothes of the very poorest on a holiday festival,—in the shape of the commonest earthenware utensils,—in the very salutation that you get from the poorest peasant in the fields.

Mahomedanism is certainly no friend to sculpture or painting, for it takes in a very literal sense the prohibition of the Jewish decalogue to "make no likeness of anything that is in heaven above, or the earth beneath, or the waters under the earth." No good Mussulman will even have his portrait taken, and geometric forms or the flowing Arabic sentences from the Koran are the proper ornaments of the orthodox mosque. But the prohibition has been relaxed at least in the case of flowers, and the exquisite carving and inlaid work of the Taj are almost unapproachable in excellence, while the fretted marble screens at Jyepore and elsewhere are more like lace than stone.

I can tell you but little of the inner life of all these millions of our fellow-subjects, for it is jealously guarded from European eyes; they and their white masters lead a separate existence, as I have already said, in all that concerns social and domestic matters, the one living in their towns and villages, the others in their stations or cantonments from one to five miles distant. But in their external life, which we *do* see, they are generally a quiet and simple race, temperate in their living, patient and much-enduring, not deficient in courage, great fatalists and very superstitious; of strong domestic affections; fond of holiday making and childish amusements. Public spirit, philanthropy, chivalrous feeling, high principle, truth, chastity and generosity,—these, as we all know, are the outgrowths, if not of Christianity, at least of a healthy national life; and in these qualities natives are, as a rule, deficient. But this at least may be said, that those amongst us who see most of and live most amongst them like them the best, and are the readiest to admit their many good qualities, and to acknowledge that, in spite of the differences created by religion, colour, race, climate, and manners, there is no such great difference, after all, in *human nature*.

And now I should like to tell you something more of Indian life in its English aspect, and make you congratulate yourselves or regret (according to your tastes) that your lot has not been cast in India.

Well, an Indian up-country Station, say in Northern India, is a piece of very flat ground, divided by very straight roads into very square patches, or "compounds" as we call them, which are bounded by prickly cactus hedges, or often by low mud walls. In each "compound," or enclosure, stands a one-storied house, with more or fewer rooms, and of greater or less size, according to the rent paid for it. Behind and near the house is a row of mud huts inhabited by your servants, a range of stables, and a kitchen or cookhouse, whose very primitive arrangements would drive an English cook into a lunatic asylum, but in which your native *chef* manages to turn out very presentable dinners.

Part of the compound is occupied by the garden, in which your Malee, or native gardener, grows vegetables, which naturally cost you a great deal more than if you bought them in the bazar. However, if you want peas and beans, you *must* grow them, for they are not to be found in the town; the natives don't eat them themselves, for neither their fathers, nor their grandfathers, nor their great-grandfathers did so, and why should they? Potatoes having been introduced some 400 years or so, are now beginning to be eaten in some parts of the country by the people, not I suspect without many misgivings lest they should be suddenly converted to Christianity by eating white men's food. Besides English vegetables grown from imported seeds, the kitchen garden produces some very nasty native vegetables, trees growing plums and peaches only fit for tarts, a bed of strawberries if you are very lucky, a patch of Indian corn, and last, not least, two or three mangoe trees, from which you get the only fruit fit to eat in Northern India, unless I except the melon.

But let us go into the house and pay our respects to the lady. A sable attendant (as the novels say) is asleep in the verandah, and is with some difficulty aroused—"Mem sahib hai?" Is the lady in? "Han, sahib,"—Yes, sir. You send in your card, for no native can be entrusted with the task of pronouncing English surnames, and a story is told of a new arrival ignorant of this fact who, desiring the native footman to announce him as the Honourable Hastings Sahib, to distinguish him from a Mr. Hastings who was *not* Honourable, was considerably disconcerted at being announced as Horrible Estink Sahib.

The man has returned and says we are to go in; so we enter and find the lady (looking very like most English ladies, only perhaps rather paler), seated in the drawing-room, the punkah swinging violently, for it is the hot weather. You see the room is a good-sized one, at least 20 feet high, with six or eight doors in it, all indispensable for ventilation; the floor covered with China matting which is cooler than a carpet; not too much furniture in it, for that would make it look hot, but a piano, books, and flowers at least, and a few pictures on the wall. The master of the house is not visible; he is either at his office, working in a room crowded with natives and the thermometer at 96°, or taking it easy in his own room, in his shirt-sleeves and slippers, smoking, and reading or writing. We sit down and talk to the lady about the dreadful heat of the weather, the chance of her going to the hills, the good looks of the last arrived young lady in the station, the dinner party at the Brigadier's the night before, and the chances of getting up private theatricals next cold weather.

If she is very civil, she may ask us to stay to tiffin, when we are regaled with salmon from England, in hermetically sealed tins, curry and rice generally very good, and Bass's pale ale, invariably termed *beer* in India; after this, we take our leave; the lady probably takes her siesta, and then in the evening goes out for a drive to the band and returns to dinner.

As to your own mode of life, if you are a sensible man, you will always rise early in the hot weather, say at five o'clock, and get a walk or ride in the cool of the morning, coming home before seven to what is called "little breakfast," where you drink tea and eat fruit if there is any, or iced mangoe fool; after which, you had better read or write till it is time for regular breakfast, and then go away to your office. Men holding official positions get a pretty good spell of desk-work every day; they are far better off than those who are not forced to work hard, for the heat is so enervating that it is very difficult to work as an amateur.

If you can get a month's holiday, of course you run up to the hills, where at an elevation of 7000 feet above the sea level, you find yourself in a charming climate and beautiful scenery. The Governor-General and all the heads of departments go up now regularly for six months out of the twelve, like sensible men, for which they get well abused by the press, the editors being obliged to stay in the hot plains. Men who can afford it send their wives and children up every hot weather, and run up there when they can get short leave. It is expensive work, but if you want to keep your children alive, there is no resource but the hills or England; they wither and die in the plains like plucked flowers.

In the cold weather, life is much better; the absentees return from the hills and society rouses itself up for a little gaiety. If you are a sportsman, you can generally get shooting of some sort, and occasionally fishing, or you may have a spell of life in tents, and go about like the patriarchs of old, literally "with your flocks and your herds and your little ones,"—for you take every single thing with you, and travel at the rapid pace of ten or twelve miles a day. The duplicate tents go on the previous night on bullock-carts, or camels, and when you ride to your new camp in the morning, you find them ready pitched under a clump of trees, and breakfast waiting; after which you set to your work as if at home. The tent you have slept in was struck when you started, and travels on during the day at the rate of a mile an hour, for bullocks won't be hurried and the roads are not over good. In the evening, the elders of the village come and pay their respects to you, chat about the state of the crops, or the amount of fever in the district, which generally results in a request to you for some medicine, especially quinine. You may go out for an hour or two into the neighbouring sugar or mustard fields, and knock over some peafowl or partridge; after which it is time for dinner, and then cold enough for a camp fire. There are few who do not enjoy camp life, for a time at least; when you have to go about alone, *i. e.* with no fellow-countryman, and may not hear your own language spoken for four or five months together, it is apt to get monotonous.

No one, as I have said, makes fortunes in India now, the time has gone by; salaries are higher than in England, of course, but the expenses are enormous. The nabob of the old novels, with a yellow face like a baboon and a dried-up liver, passionate in temper, telling impossible tiger stories, and suddenly turning up in England with a hookah, two or three black servants, and several lakhs of rupees for the hero and heroine—is a thing of the past. His successor was dear old Colonel Newcome in fiction, or better still, Henry and John Lawrence, James Outram, Herbert Edwardes, John Nicholson, Henry Durand, and men of that stamp, in reality, wise in council, resolute in action, God-fearing always, with duty ever present before them as the one motive of their lives; strong men to whom both their countrymen and native subjects looked up, as worthy to rule and to be obeyed.

It may be useful if I say a word as to the cost of living in India; this has, doubtless, greatly increased within the last ten years, but of course admits of very varied estimate according to a man's tastes and idiosyncrasies. I *have* known a subaltern live on 150 rupees a month, which may be set down as the smallest possible sum for a single man, and if you double that sum a young man should live comfortably on it, without being able to indulge in any extravagancies. Many young married couples live on this, but they must exercise great self-denial to do so, and this of course is for current expenditure, and makes no allowance (in their case at least) for first cost of furniture, horses, travelling expenses, &c. The unit of coinage is the rupee, which, though nominally worth 2s., is practically given wherever you would in England give 1s., or even sometimes 6d. But, although this is the case, there is some compensating advantage in the fact, that there is not much temptation to extravagance in India, and that social life in the up-country stations is on an easy and natural footing that prevents any foolish ostentation on the one hand, or on the other, any inability to mix in society on the ground of not being able to afford it. This arises from two causes: First, that few men in India have private fortunes, and anyone may know the amount of his neighbour's income by simple reference to the pay code. Secondly, that the society, except in a few large places, is so limited that everyone who can make himself or herself agreeable is welcome on that account, and can enjoy hospitality freely without an uneasy sense of obligation arising from inability to return it in kind. On the whole, I think Anglo-Indian society is pleasant and agreeable; few men suffer from *ennui*, or think everything a bore; most of the men are interested in their work and talk a certain amount of shop—not more so, I think, than other men; there are plenty of book clubs at all stations, and most people read a good deal; there are few stations at which sport of some kind is not procurable for those who have leisure for it; and those who can get two or three months' leave to travel in Cashmere or the Himalayahs will certainly not sigh for Switzerland or the Alps.

While I am on the subject of society, let me disabuse you of the notion still prevailing in the minds of many respectable people, as well as second-rate novelists, that ladies in India are different from their English sisters, and that, when young, they are shipped over to the East like figs, to the market of Hymen. Such fables belong only to the time when every Indian was a nabob, with no liver to speak of, and the complexion of a China orange. Young ladies now go to India, I assure you, for the same reason that other young ladies stay in England,—because their fathers and mothers are there; and stray females, bent on an independent matrimonial cruise, are, alas! things of the past. On the other hand, while the one terrible drawback of Anglo-Indian life is the continual separation of husband and wife, parents and children, I am bold enough to say that those social relations—the former, at any rate—are stronger even than in England, if only on this account, that each is less independent of the other than at home. If Damon and Phyllis have had a quarrel, Damon can't go to his club, nor Phyllis run to her mamma, to make matters worse; and when each has cooled or sulked a little, Damon remembers the terrible time when they said good-bye to their little ones at Southampton, and Phyllis recalls the intense pleasure of again meeting her husband, after that cruel illness which sent her home when he could not go with her; so they kiss and make friends, and resolve to keep their tempers in better check in future.

I have often been asked how we travel in India? Well, where there are no railways, and you have a good metalled road, you travel by horse-dâk, as it is termed;—that is, you hire a carriage, which is a four-wheeled cab slightly altered, and engage relays of horses at every five or six miles; and by travelling all night to escape the heat, you manage to accomplish

60 or 70 miles between dinner and breakfast-time pretty comfortably, at a cost of about a shilling per mile.

The horses are an abominable set of brutes, and generally begin by obstinately refusing to stir a step, or, perhaps, lying down in the shafts. Then the native coachman, after patiently undoing the rotten harness and getting the animal once more on his legs, addresses him by all manner of endearing epithets. "Go on, my brother," (horse doesn't move); "go on, my son, my brave, my hero," (no result); a cut with the whip, followed by a vicious kick from the horse. Coachee changes his language—"Go on, you scoundrel—you villain;" a shower of blows, a volley of abuse from the passenger inside, more kicks and plunges from the horse, and finally, by the aid of the whole stable establishment, who push behind, whack the horse, and simultaneously yell like fiends, off you go at the rate of ten miles an hour, the horse never stopping until he gets to the end of his stage.

If there is no road, as happens very often, you must be content with the old-fashioned palanquin or doolie, a species of coffin with doors at the sides, in which you are carried on men's shoulders at an average rate of three miles per hour, the bearers being changed every ten miles. When I first landed in India, I travelled from Calcutta to Allahabad, a distance of 500 miles, in this way, taking twelve days on the road; the journey being now accomplished by rail in about twenty hours.

After your night's journey, in which you have probably accomplished some 35 miles, you are glad to espy a solitary house—a dâk-bungalow, as it is termed—in which you can get temporary rest and refreshment; so the doolie is deposited in the verandah; the hot and dusty traveller, who has hardly had a wink of sleep all night, emerges from the inside of the coffin, and summons the native servant, who in white robes and with a venerable grey beard, might sit for a portrait of any of the Old Testament patriarchs. Then ensues something like the following dialogue, which I must translate from the Hindustani, in which it is spoken:—"What can I have for breakfast?" say you.—"Your lordship can have everything."—"Very well, then, bring me a beef-steak."—"Beefy-e-steak! nourisher of the poor! how can your slave get any?"—"Well, then, a mutton chop."—"Mutton-e-chăp! cherisher of the humble! there are none left."—"Well, what *is* there?"—"Perhaps, O great king, there may be a fowl."—"Then get it ready." Exit the servant, and in another minute you hear a violent *clucking* in the back-yard, and see the man in hot chase of a venerable cock, who evinces the strongest disinclination to have his few remaining days curtailed to satisfy your unholy hunger. But it is all of no use—he is captured; his head is turned towards Mecca, and the Mahomedan cook, muttering "Bismillah," "in the name of God the compassionate and merciful," cuts the fowl's throat, and in twenty minutes more it is smoking on your table.

And now my watch warns me that I have tried your patience long enough. I have endeavoured to give you some idea of the nature of Anglo-Indian life, and of that country which has so often been described as the brightest jewel in the British Crown. But, after all, the only way to know much of a country is to see it; and I often wonder that more Englishmen do not pay a flying visit to India now that it is so accessible. Many certainly do go there now-a-days for a short time, and are, I am sure, well rewarded for their pains; but they are very, very few in proportion to those who might go, if they only knew how to set about it, and perhaps I cannot conclude my lecture better than by taking you all a journey there on paper, just as if I were Mr. Cook, and you were a party of tourists.

We start from the Waterloo Station and run down to Southampton, going on board one of the fine steamers of the P. & O., as it is always called, which I need not say is short for the Peninsular and Oriental Steam Navigation Company. The mails are on board, the last good-byes have been said, and we are steaming down Channel and across the Bay of Biscay, until, on the fifth day, we cast anchor off the Rock of Gibraltar. A few hours' delay for coaling, and we run on to Malta, and find ourselves in Valetta Harbour, wandering about the quaint old town, eating oranges, buying lace, or examining the old church of St. John, with its relics and monuments that carry us back to the middle ages. Another three days' run and we are landed at Alexandria, and getting our first peep at the East from the top of a donkey. But the train is ready, and we speed through the flat, fertile country watered by the Nile, cross that muddy and venerable stream which has puzzled all the geographers from Herodotus to Livingstone, get a glimpse of the Pyramids in the distance, and cross that terrible desert which is still what it was when Moses led the Israelites through it more than 3000 years ago; and in twelve hours from the time of leaving the Mediterranean, are on board the other mail steamer in the Red Sea. Six days' run takes us to its mouth, and we are anchored off Aden, a strong military post built on a barren rock, where we stay a few hours for coaling, and then enter on the last stage of our voyage. Five days' run across the Indian Ocean and we sight the magnificent harbour of Bombay; and then a little steamer carries us all off to the shore, and we are driving to Pallonjee's Hotel under a fierce sun, tempered only by the sea-breeze, and through streets thronged by motley crowds of natives, in which the few white men are altogether lost.

From Bombay, the Great Indian Peninsular Railway will take you 600 miles in about twenty-five hours to Jubbulpore, up the steep inclines of the Western Ghat mountains, and through the dense jungles and undulating hills of Central India.

Another run of 220 miles takes you to Allahabad, the modern capital of the North-west Provinces, situated at the confluence of the Ganges and Jumna, a great military and civil station, but dusty and disagreeable. Here resides the Lieutenant-Governor of the

North-west Provinces, a territory larger than France, with 30,000,000 inhabitants. From Allahabad we may travel south-east to Calcutta, 600 miles distant, passing Benares, the holy city; Patna, famous for opium; and many others; or north-west to Agra, Delhi, and Lahore; the last, 700 miles distant, the capital of the Punjab, where resides another of our great pro-consuls, as Macaulay calls them, *viz.* the Lieutenant-Governor of the Punjab.

All these great cities are well worth a visit, containing, as they do, many interesting ruins and architectural remains of the old Hindoo and Mahomedan rulers of the country. Beyond Lahore, where the railway at present terminates, a fine road 260 miles long, will take the traveller across the classic streams of the Punjab, famous for Alexander the Great's campaigns, up to Peshawur, on the extreme north-west boundary of the empire, where a force of 8000 men keeps watch over the fierce and turbulent races of the neighbouring mountains.

From Umballa, on the Punjab Railway, if the traveller strikes to the east, a good road of 40 miles will take him to the foot of the Himalayahs, whence he can ascend to Simla, situated amongst its groves of pines and rhododendrons, 8000 feet above the sea; and thence through some of the finest mountain scenery in the world, to Cashmere or Thibet, or the frontier of China.

You may vary the homeward route by visiting Calcutta, Madras, and Ceylon, or by going down the Indus to Kurrachee, and thence back to Bombay.

And over all these distances, and through this vast country, the traveller may journey as safely as in any part of Europe, in a healthy, enjoyable climate, if he chooses the proper time, and with his mind expanded by the contemplation of scenery differing widely from anything in the West, and of a state of social and national life which the most superficial glance will assure him is utterly foreign to all his previous experiences.

That, I think, is the chief good of all travel. The experience it brings lifts us out of our old grooves of thought, widens our narrow ideas, teaches us that as God has not made us all with the same coloured skins, so he has given us varieties of national character, which are admirable from their very diversity, and do not make us the less members of one common family, in which He is the great Father of us all.

For purposes of administration and government India is divided as follows:—The Viceroy and Governor-General of course rules over all, making his headquarters at Calcutta in the cold weather, and at Simla in the summer. Under him are the Governors of Madras and Bombay, the Lieutenant-Governors of Bengal, the North-west Provinces and the Punjab, with their seats of government at Calcutta, Allahabad, and Lahore respectively; the Chief Commissioners of Oudh, the Central Provinces, Central India, and British Burmah, residing at Lucknow, Nagpore, Indore, and Rangoon; and the Commissioners of Mysore, Hyderabad, and Scinde. Some idea of the extent and importance of these provincial commands may be formed if we consider that Bengal proper, for example, is as large as Germany, and numbers 60 millions of inhabitants.

With regard to the Public Works of the country, their control and direction are confided to a separate department of the State, known as the Public Works Department, at the head of which is a Secretary, styled the Secretary to the Government of India in the P. W. D., and who is, in effect, the consulting engineer and professional adviser to the Viceroy and his council. He is assisted by Deputy-Secretaries for the separate branches of Irrigation, Railways, and Military Buildings.

To the head-quarters of each local government is similarly attached a Secretary, who is at once the mouthpiece and professional adviser of the Lieutenant-Governor or Chief Commissioner, and who, as Chief Engineer, is also head of the whole public works establishment of his province.

Subordinate to him are the Superintending Engineers, who may either have general charge of all the public works of a large district, or, as is more generally the case, are in special charge of some large work, such as a Canal or a Railway. Under the superintending engineers are the Executive Engineers, who are, in effect, the working units of the system. An executive engineer may have a range of new Barracks to build, a line of Road 80 to 100 miles long to keep in repair, or 30 or 40 miles of Canal or Railway to lay out and construct, and for the actual execution of this work he is primarily the responsible man. If the work is a new one, he has to prepare the detailed designs and working drawings under the advice and guidance of his superintendent, to frame the estimates, and to write the reports. When sanctioned, he has to lay out the work, and to find the workmen or contractors to execute it, to control the expenditure, to submit monthly accounts and progress reports, and to conduct a tolerably large correspondence. He will probably have two or three Assistant-engineers, five or six European Overseers, and eight or ten native Sub-Overseers, besides an Office establishment of clerks and accountants.

All the above officials, from the grade of Chief Engineer down to that of Assistant-Engineer, inclusive, are indifferently drawn from officers of the Royal Engineers; Artillery or Line Officers trained at Roorkee; Civil Engineers sent out from England; civilians (European and Native) trained at Roorkee. The promotion from one grade to another is partly by merit and partly by seniority, and has nothing whatever to do with military rank; but care has generally been taken not to have a Royal Engineer officer serving departmentally under his junior.

It may interest some of you if I trace the probable career of a young Royal Engineer, sent to India, say, to what is still called (though improperly) the Bengal Presidency. On landing at Bombay, and reporting himself to the military authorities, he will be directed to proceed to Roorkee, where he will have to do duty with the Bengal Sappers and Miners for a year. This is the order at present, the idea being that in the interval he will acquire some knowledge of the language and of the customs of the country, and, if he is wise, he will make good use of his time; for until he can speak Hindustani pretty fluently, he will find himself very helpless, and all but useless. Roorkee is the head-quarters of the Bengal Sappers, and virtually the head-quarters of the Royal Engineers in the Bengal Presidency. The college over which I have, for some years, had the honour of presiding is at the same station, but is in no way connected with the Sappers, being a public works institution under the civil government, while the Sappers are of course under the Commander-in-Chief.

The Thomason College has been founded about twenty-five years, and now contains about 250 students. There is an engineer class, consisting of a few Artillery and Line officers, and some thirty civilians, who undergo a two years' training to fit them for the posts of Assistant-Engineers in the P. W. Dept.; another class of officers, who stay only seven months, and are trained for the Quartermaster-General's Department; a class of soldiers who are trained as Overseers; and a large native class, who are educated as Sub-Overseers, Sub-Surveyors, Estimators, and Draftsmen. Besides the Principal, there are two Royal Engineer Assistants on the staff, two civil Professors of Mathematics and

Experimental Science, and two sets of subordinate Masters for the lower classes. There are also a library, model room, and museums in the college, and an excellent press, whence a good many useful works have issued, chiefly relating to Indian engineering.

The Bengal Sappers and Miners are a fine body of men, consisting of twelve companies of native soldiers, recruited from the best and most warlike tribes in Upper India, who have done excellent service wherever they have been employed. Besides their own native officers, English non-commissioned officers are attached to the companies. There is a Commandant, Adjutant, Superintendent of Instruction, and Superintendent of the Park and Field Train, and four doing-duty officers, who are all Royal Engineers, besides the new arrivals attached temporarily to the corps. About half the men are at head-quarters, the remainder in detached companies at Peshawur and elsewhere on the frontier. There is a very fair park and pontoon train attached; also workshops and schools. The men are skilful and intelligent, excellent workmen and good soldiers.

Roorkee also possesses a Foundry and Workshops belonging to Government, which are interesting as having been the first of the kind erected in India, twenty-four years ago, before the introduction of railways. The workmen are all natives, and some of them are remarkably clever and intelligent. They will make anything for anybody, from an iron bridge or a steam engine, down to a railway key; and they turn out excellent spirit levels, prismatic compasses, and so forth. Near Roorkee are also all the greatest works of the Ganges Canal.

Roorkee is a pleasant and healthy station, and many of the young officers stay voluntarily with the Sappers for more than the regular year, as their departmental promotion counts all the same. But generally speaking, before the year is out, the young officer will read in the Gazette of India one fine morning that his services have been placed at the disposal of the P. W. Dept.; and in the next Gazette that he is posted to such or such a province; then a week later, in the local Gazette of that province, he will be posted to a particular circle, and the superintending engineer of that circle will desire him to report himself to some particular executive engineer. The day after his arrival he will find himself employed according to the nature of the work, either surveying and levelling, or drawing plans and making calculations, or in a tent in the middle of the jungles superintending the building of a bridge, with not a soul that can speak a word of English within 30 miles of him. For the next four or five years he will probably be changed about a good deal from one work to another; and if he has proved himself efficient, will then find himself an executive engineer of the fourth grade, and in charge of a division; while, after running through the four executive grades, another ten years or so may carry him on to the higher grade of a superintending engineer.

Generally speaking, a new arrival can select his own line of departmental service, which will depend on his tastes and circumstances. If he has been weak enough to get married before going out, or if he is fond of society, he will select the Military Buildings' branch, so that he may live in a station; and if he has a speciality for architecture, I should strongly recommend him to do this, for there is a great want of men in that line in India. If, however, he is fond of hard work and knocking about in the jungles, and doesn't mind a solitary life, he will prefer the Irrigation or Railway branch, and I do not think he will regret his choice, for no man fond of his profession could desire more interesting work, and in the construction of a new line of Canal or Railroad he will every day find scope for his talent or ingenuity or readiness of resource.

Not a few men enter the Survey, which is, however, quite a separate department, divided into three branches,—the Trigonometrical, Topographical, and Revenue,—and the promotion in it proceeds *pari passu* with that of the Public Works Department.

It may be asked, what are the relations of a Royal Engineer officer, in a department so miscellaneously constituted, to the Civil Engineers and Line officers, with whom he has to work? I think they are generally very friendly and agreeable, and though, of course, it is pleasanter to serve under one's own brother officers, yet the various members of the department work harmoniously together, and the promotions are generally very fairly made, energetic and clever men being pushed on well.

You will observe that I have said nothing about military engineering works except barracks; in fact, there is very little work of that kind in India. Like the Romans of old, we encamp our troops in open cantonments instead of shutting them up in forts, and what forts there are in the country, for the magazines and arsenals, are almost all old native forts. Of harbour defence works we have scarcely any, for the simple reason that there are scarcely any harbours, except at Bombay, where the subject has recently had much attention directed to it. Whatever military engineering work is required is therefore done as a civil work by the Public Works Department, and all Royal Engineer officers virtually work as civilians, except in the event of war, when those required are at once ordered off to join the army; and though doubtless an officer's knowledge of military details may be considerably weakened by his long absence from military duty in civil employment, yet the very nature of that employment engenders a quickness and fertility of resource and a sense of responsibility which go far to compensate for that deficiency, and the men who blew in the Cashmere Gate at Delhi, and laid out the defences of the Lucknow Residency, at any rate found their knowledge of military engineering sufficient for the purpose.

What I have already said as to the importance of the work entrusted to every young Government civilian in India applies almost as much to the young Engineer. He will find himself almost immediately entrusted with responsibility, and, before long, in charge of work that he could not expect to have confided to him in England until many years of service had rolled over his head. But with work and responsibility of this nature, he will find great interest and no small anxiety; he will have to look after or do nearly everything himself, without the aid of clever contractors, skilled clerks of the works, and intelligent foremen; he will probably have to train his own subordinates, to work with very inefficient plant, and to trust to his own resources every day, and under circumstances calculated to try his mother-wit, his common sense, and clear-headedness, above all, his patience and temper, in a way he never calculated on. No sort of knowledge will come amiss to him; he may even have to be his own doctor, and to physic his whole establishment, or to commence work by turning wholesale provision merchant in order to supply his workmen in some barren district.

Indeed, the work of an Engineer officer in India, as in England, is of a very miscellaneous description, and I think it is no light subject of pride to the whole corps that there is such a variety of talent amongst our brother officers. In India, when I left the country, besides Lord Napier, whose reputation belongs to the empire, we had the ablest Lieutenant Governor in the late Sir H. Durand, who began his career by blowing in the gates of Ghuznee. One of our officers is Director of Indian Telegraphs another superintends the Great Trigonometrical Survey, perhaps the greatest and most scientific survey ever yet achieved; another is head of the Calcutta Mint; others were amongst the ablest civil commissioners of the country; the astronomical observations of another in connection with the last two eclipses have been of the highest value to science; another has just received the founder's gold medal from the Royal Geographical Society for his works on antiquarian geography; another has perished a martyr to science amongst the deserts of Thibet; another has established a reputation as the best accountant in India, and has since written the most famous pamphlet of the age.

To complete my account of the Public Work agency, I must very briefly refer to some of the financial aspects of the Department, which concern the engineer. How is the money found for the various works of the country? In this way:—

At the beginning of the financial year, the financial department of the Government of India allots to each local government the several sums that are authorized to be expended during the coming year, for the several works that may have been sanctioned under the various heads of irrigation, railways, roads, barracks, &c. These sums, thus provided in the Public Works Budget, are allotted after due consideration of the various budget estimates that have previously been submitted by the chief engineers of the local governments, and they are not allowed to be exceeded. For this, the superintending and executive engineers are held responsible, and they must "cut their coat according to their cloth," *i. e.* arrange for the coming year's work according to the amount of money they are authorized to expend.

Every work is estimated for previous to sanction by the executive engineer, and the estimate, after being checked by the superintending engineer, is forwarded to his chief for sanction, who, if he approve the design and estimate, recommends it for sanction by the local government, or, in case of a large work, forwards it on to the supreme government with his own remarks. If not satisfied with it, he may return it for revision or explanation.

When a work is sanctioned and ordered to be commenced, the money being also forthcoming for it in the budget, the executive engineer goes to work. In the large presidency towns, and a few of the more important stations, he may get the work executed by contract; but, as a general rule, he will have to be content with procuring his materials by contract, and perhaps getting his earthwork done in this manner. For the rest, he will have to employ daily paid labourers, and occasionally may have to import labourers from other districts, to organize them into gangs, provide them with tools, and arrange for their food, water, and temporary shelter. For this he has the assistance of his European overseers, native sub-overseers, and *mistrees* or head-artificers, and as you will have to deal largely with these men, it may be as well if I say a few words about them.

The European Overseers are nearly all non-commissioned officers or privates who have volunteered from the various regiments in India for the Public Works Department, and have been trained at the Roorkee College. They are allowed to wear plain clothes, and are of course struck off all military duty. As a rule, they are hard-working, intelligent men, and many of them are most valuable subordinates, but they are generally deficient in practical knowledge, are not very conversant with the language, and are but too often given to drink.

The native Sub-Overseers have also been trained at Roorkee, and are generally good draftsmen, surveyors, and estimators, but they are drawn from the trading instead of the working classes, have no practical experience, and lack physical stamina.

The Mistrees, or native head-masons and carpenters, are generally intelligent and good men, quick to learn and easily managed, but few have any theoretical knowledge.

The native labourer is patient, docile, and lazy, never drinks, and is easily managed by anyone who understands him. Perhaps this is a good place to say a word or two about the natives generally, and their treatment by their English masters. Those who have any inherent antipathy to black or brown skins had better not go to India, and those who do go, and are anxious to find faults in the natives, will have no difficulty in satisfying themselves on that score. But, as I have already said, those Englishmen who live most amongst them, and have most to do with them, get to like them most, with scarcely an exception, and I can honestly say, after twenty years' experience, that I am no exception to the rule. Learn their language well, spend a little time in studying their habits, prejudices, and modes of thought, and I am sure you will find the trouble repaid. If they are not very truthful, are indolent, and sometimes troublesome or even exasperating, it is no light thing that they are singularly temperate, wonderfully patient and good-tempered, very susceptible to kind treatment and good management, and that strikes, drunken brawls, and grumbling discontent are simply unknown.

Now, after so much preliminary dissertation, we may come to the more professional part of our subject, and I shall begin by making some remarks on the Materials with which the Indian engineer has to work, after which we will proceed to discuss the particular constructions in which they are employed, so far as these present any special points of interest to those conversant with similar structures at home.

There are many varieties of *Stone* in different parts of India, and it is employed in the various forms of ashlar, rubble, &c., very much as it is in Europe. Granites, limestones, and sandstones are extensively used in the localities where they occur, but the cost of carriage over bad roads to distant places necessarily restricts the employment of this material. In Southern India, laterite, a clay-stone, is extensively used, being easily worked, and becoming hard by exposure to the air. In Upper India, Delhi and Agra are famous for their red sandstone, and Jyepore for its white marble, of which the Taj and other famous buildings are constructed. Bombay has also many varieties of stone, notably the Poree-bunder limestone. Allahabad has some fine quarries of sandstone, of which the new government buildings have been constructed, and I recommend to your notice the account of the working of the Purtarpore Quarries, in the Professional Papers on Indian Engineering, as giving much practical information. Slate is generally scarce and inferior, but some fine quarries have lately been opened out at Dalhousie, and in the Khuttuk Hills, in the Punjab.

There is a kind of soft stone called moorum, found in Central and Western India, which, though almost useless as a building material, is extensively employed for road metalling. Kunkur, too, is quite an Indian speciality, though it is almost entirely confined to the North-Western Provinces. It is a peculiar kind of oolitic limestone, found in beds just below the surface, and is of two kinds; one adapted for building purposes, in which it strongly resembles artificial concrete; the other answering admirably for road metalling, for which purpose it is broken into lumps about the size of an egg, drenched with water, and then rammed until perfectly smooth, after which it is allowed to dry before the traffic comes on it.

The manufacture of artificial stone by Ransome's process has been tried at Bombay on a small scale, but not with success in an economical point of view.

In the greater part of Upper India, and over much of the rest of India as well, *Brick* is the chief building material, and there are few engineers in India who will not have much to do with brick-making. I would therefore strongly recommend you to make yourselves well acquainted with the latest improvements in the art, at the same time bearing in mind certain Indian specialities, which will limit the use of many of these methods; these are the cost of carriage, the general absence of coal fuel, the dearness of other fuel, the absence of skilled subordinates, and the disinclination of natives to be driven out of their own customs, and to try experiments. But these circumstances, remember, should only serve to guide your inquiries, on no account to restrain them, for similar difficulties will always be found in the path of improvement everywhere.

You will find very full information on Indian brick-making in the 'Roorkee Treatise on Civil Engineering,' and I shall only here advert to a few salient points of the subject.

There is plenty of good brick-earth to be found, but the cost of carriage prevents the same care being taken as at home in the selection and admixture of clays. The clay is often tempered by hand, and then taken straight to the moulding-table; but pug-mills are now pretty common, worked by bullock power. The bricks are usually sand-moulded, and are made of the English size, and stacked in temporary sheds to dry. Brick-making machines have occasionally been tried; but their expense, the difficulty of repair, and the cheapness of hand labour, have always driven them out of the field. Hollow bricks, too, are never seen; and as I think they would be found to be much cooler in the walls of buildings than solid bricks, I would recommend anyone to acquire information of their manufacture and cost.

Bricks are burned in clamps and kilns as in England; but it is only in the neighbourhood of the very few coal localities in India that coal fuel can be used; you will generally have to use wood, or in the case of clamps, dried cow-dung and stable litter, commonly called *oopla*.

At Akra, near Calcutta, there is an extensive Government brickfield, which I advise those of you who are able, to visit; Mr. Hickmott, the superintendent, is a very civil and intelligent man. Hoffman's kilns have been lately tried, but not very successfully; and I advise you to study the subject, economy of fuel being even more important in India than in England. At the same time, the first cost of construction in this, as in many other instances, must be clearly kept in view; for this cost will have to be added to the cost of your bricks, and as it would not pay to carry bricks far, and Indian distances are long, you cannot afford an expensive construction, however good in itself, which would all fall on the cost of perhaps two or three hundred thousand bricks.

Coloured bricks are nowhere used in India, and their absence is much to be regretted, for they would be most useful both for architectural ornamentation and for floors and similar purposes. The proper clays, on which, as you know, the colours depend, are found in some parts of India; and careful search would doubtless bring to light others;—but here again we are met by the fact that their manufacture requires skill and capital, which are not found on the spot, and would have to be imported. The same remark applies to terra-cotta and encaustic tiles, which would be admirably adapted for Indian use, and would, moreover, stand much heavier transit charges. But the Government cannot be expected to enter the field as manufacturers, and so we must wait, I suppose, for English capital and skill, or for the progress of native enlightenment on these points.

Good ordinary bricks are, however, generally procurable in India, if only proper care be exercised, and a fair price paid for them. The bricklayers require close watching, and often systematic instruction in the all-important subject of bond; for the common native brick is very small, and laid in quantities of mortar with little care about bond; so that native walls are really masses of concrete.

Brick arches are laid, as in England, either in half-brick rings, or, in important works, with the bond carried right through the arch. The natives are very skilful in constructing cheap centerings of dry bricks and wooden soffits, plastered over with mud to the form of the arch, which answer well when your centering is not liable to be carried away by a rush of water. For large arches, whether built with regularly framed timber centres or the common native centres, the French fashion of striking them by means of hollow iron cylinders filled with dry sand, supporting pistons on which the laggings rest, has been largely employed of late, and is much to be commended. You will find it described in Nos. 11 and 24 of the 'Roorkee Professional Papers,' as also in the 'Treatise.'

Masonry made of brick or stone laid in lime mortar is everywhere called *pucka* masonry. Bricks are, however, often laid in a mortar made simply of thin clay with a little chopped straw in it, and the work is then *kucha-pucka* masonry. Sometimes sun-dried or unburnt bricks are used, in the case of cheap buildings, for interior walls; or in districts where the rainfall is scant, for exterior walls as well; this is *kucha* masonry. If well executed, and covered with a leping of clay and cow-dung,—the foundations, tops of walls, and timber framings being finished with pucka masonry—this cheap kind of work answers well in very dry districts, and is cooler than burnt brick.

*Tiles* are also extensively used in India for roofing purposes, and they are often very badly made. I cannot, however, do more than draw attention to the fact here. You will find plenty of information on the subject in what was once described to me, the author, as "that refreshing work," the 'Roorkee Treatise on Civil Engineering.' The tiles generally used in Government buildings are known as the Goodwyn and Atkinson pattern tile. The hexagonal hollow tile introduced by Colonel Fife for roofing purposes (and known as the Sindh tile), is worth your attention; also the drainage tiles made by Captain Jeffreys for the Ganges Canal. These I mention to illustrate what I have already told you of the necessity of engineers becoming their own manufacturers if they wish to make improvements.

We come now to the subject of *Limes* and *Cements*, a very important one in India. Lime is obtained in India from the limestone boulders found in hill torrents, from kunkur, from beds of marl, or rather calcareous tufa, and from limestone *in situ*. It is burnt with wood fuel, sometimes in the open, generally in conical kilns, and is mixed with sand, burnt clay, or brickdust, and sometimes other ingredients, to form mortar in the usual way. The best lime is that procured from boulders, which when mixed in the proportion of one part lime to two parts of *soorkee*, or pounded brick, forms an excellent mortar for hydraulic works. Kunkur lime, as a rule, is simply mixed with sand. When lime is burnt with *oopla*, care must be used in sifting and separating it from the ashes of the burnt fuel, otherwise, of course, its strength will be greatly impaired. Artificial cements have scarcely hitherto been made or used in India. You will find a valuable paper on the relative cost of the manufacture and importation of Portland cement, in No. 294 of the 'Professional Papers'; also another paper by Mr. Price, describing the manufacture of artificial hydraulic lime at Kurrachee. There can be no doubt of the feasibility of the process; but as a rule, the mortars employed, in Upper India, at least, are excellent, if only proper care be used in their preparation. What is chiefly required is a very quick-setting cement or mortar which, even when used in building in water, shall harden in one or two days. This is much wanted for repairs to canal works, where it is often absolutely necessary to turn on the water before the mortar generally employed has had time to set.

Concrete is not very much employed in India, though it has attracted a good deal of attention lately; and some of the works on the new Sirhind Canal were designed to be built almost entirely of it, such as arches of 40 feet span. Indeed, with an abundance of excellent lime, and a great scarcity of fuel, it seems curious that it has not been more extensively used. Concrete blocks were recommended for the great weir over the Ganges, by the Ganges Canal Committee, to be composed of shingle, sand, and lime; and if proper apparatus be used for testing the quality of the lime, there seems every reason to anticipate economy and good work from such a mode of construction.

Lime is also used in stuccoes and plasters much as in England. Madras is noted for this work, where the very beautiful *chunam* plaster for interiors of rooms, is as smooth, hard, and polished as marble. Coarse sugar and pounded egg-shells are mixed with these more expensive plasters.

Of *Timbers* there is an immense number in India occasionally used; but practically you will find yourself restricted to a very few varieties, which are the only ones procurable in any quantity. In the Punjab, for instance, the Deodar (Cedrus Deodara) is the principal wood employed, being nearly identical with the famous cedar of Lebanon. It is found in the Himalayan forests, where it is cut, thrown into the rivers, and left there till the succeeding rains swell the stream and carry the logs down below, when it is rafted and floated into the plains. It is a very valuable timber, procurable in great scantlings, and used for every purpose: trees of 7, 8, or 9 feet diameter at the foot, and 70 feet long, are by no means uncommon. In the North-Western Provinces the Saul (Shorea robusta) is the principal tree; it has a long fibrous grain, is straight and strong, of a reddish colour, and very valuable for all purposes.

In Burmah and Western India, the Teak is the principal wood: its many excellent qualities are doubtless well known to you. Other common timbers are the Mangoe, used only for planking or furniture, and readily attacked by insects; the Sissoo or Sheeshum, a hard, strong, but crooked wood, in general request for many purposes, especially furniture, as it takes a beautiful polish; the Keekur or Babool, an acacia, a very hard tough wood, much used for carts; the famous Bombay Black-wood, of which some beautiful specimens of carved furniture are to be seen in the Indian Court of the International Exhibition; the Toon, an inferior mahogany; the Sandal wood, which has a strong perfume, and many others.

Timber in India is generally seasoned by the air or water process, and is occasionally Kyanized or Burnettized. Well-seasoned timber stands the climate well if carefully protected from white ants, those pests of the East. For this purpose, the ends of beams fitting into walls are generally charred and tarred, or the timber is soaked in a solution of sulphate of copper; but the best preservative is carefully to prevent any earth or mud from coming in contact with it. Wooden posts buried in the earth will very soon be useless.

I don't know that there is anything special to be said about the Indian Carpenters, who are generally very fair, and sometimes very clever workmen, though they *do* squat on the ground, and hold a piece of wood with their toes while they work the drill by means of a bow and string with their hands. Carpenters' benches were introduced in the railway workshops at Lahore; but the superintendent told me that he had no sooner turned his back than the men at once proceeded to squat on the benches, so he gave them up in despair.

Of the *Metals* used by the Indian engineer I need only mention Iron, which is nearly all brought as pig from England. There are valuable iron, copper, and other ores in India; but the great cost of fuel and of carriage have hitherto prevented their being worked extensively. A good deal of native iron is certainly brought into the market, and worked up into tools, straps, bolts, &c.; but nearly all iron roofs and bridges are imported from England; even those made at the Roorkee workshops being manufactured from English iron.

Having now touched very briefly on the chief materials employed, I may usefully give you some information with regard to *Wages* of workmen and *Cost* of work. Wages, of course, vary more or less; but the pay of a common labourer all over India may be fairly set down at 2 annas, or 3*d*. a day, and of an ordinary mechanic at 6*d*. to 7½*d*., with which he finds himself in everything. A *beldar,* or navvy, will get 4½*d*., and a skilled carpenter or mason from 9*d*. to 1*s*. These wages seem very low compared with English prices; but you must remember that the men all do far less work than an Englishman; thus, the lowest estimated rate of common earthwork is now about 5*s*. per 1000 cubic feet, which is at the rate of 50 cubic feet only per day for each man of the gang employed.

Of course, I need not trouble you with a long string of rates; I only wish to give you some idea of relative prices and money value. Good ordinary brickwork will cost about 40*s*. per 100 cubic feet; ashlar about 2*s*. per cubic foot; timber-work, 7*s*. per foot, "wrought and put up."

Taking into consideration the price of food, and other things, we may fairly reckon, I think, the difference in the value of money employed in constructing public works in India and in England as 1 to 4, *i.e.* that a work costing 10,000*l*. in India would cost 40,000*l*. in England.

The Rupee, as you know, is nominally 2*s*., but really varies according to the rate of exchange between 1*s*. 9*d*. and 1*s*. 11*d*. It is divided into 16 *annas*—a copper coin worth about 1½d., which is again divided into 12 *pies*, 4 of which make a *pice*. The 2-anna, 4-anna, and 8-anna bits correspond to our 3d., 6d., and 1s. There is no standard gold coinage yet in India, though sovereigns will always pass current, and 5 and 10 rupee coins have lately been issued. There are now Government notes from 10 to 1000 rupees, which pass in all the great towns, but will not be taken in smaller places without a discount.

As to Measures and Weights, in Upper and Western India we employ the lineal, square, or cubic foot as the unit for our estimates and calculations; in Madras, they use the cubic yard. Our lineal units are generally understood, except, perhaps, the mile; natives generally give distances in *kosses*, which vary from 1 to 2½ miles. The acre is employed in Government papers, but is not known to the people, who measure land by the *beegah*, which varies greatly in different districts. The weights commonly used are the *maund*, *seer*, and *chittack*, which are practically held to be 80 lbs. avoirdupois, 2 lbs., and 2 oz. respectively.

There is as great variety in India as in other countries in all local weights and measures, and the people are as obstinate and suspicious there as in England, and elsewhere, about any changes being made,—a fact too often forgotten by the decimal and metrical doctrinaires in their zeal to ensure a theoretically perfect system of standards.

Let me now say something of the mode of executing

work generally, before going on to the subject of special engineering constructions. The first thing that strikes the engineer from England is the primitive simplicity of the working appliances, and the total absence of the elaborate and costly plant judged necessary for executing great works in Europe. Steam engines, steam cranes, steam pumps, steam pile-drivers, tramways, even such things as hand-pumps, horses, carts, and wheelbarrows are rarely used. This has arisen from the comparative cheapness of manual labour, from the difficulty of procuring skilled subordinates, and from the dearness of fuel. Now that the first two causes are greatly modified, and the extensive works lately undertaken have necessitated recourse to European appliances, the engineer going out to India should devote special attention to this subject, for he will often have to teach his subordinates the use of such things as I have mentioned. Great works have, however, been constructed in India without them, and will still be so for some time to come. Earthwork, for instance, is constructed almost entirely with wicker baskets as the sole means of carriage; yet few countries have had so many massive embankments thrown up. Petty contracts are readily taken for this kind of work, in which case the whole family of the workman, down to the child of three years old, will help to swell the mass to be raised. A system of payment at the rate of a *cowrie* paid on the spot for every basket of earth carried is very popular, several hundred cowries going to a rupee. For such banks, the earth is almost always taken from side cuttings to save the expense of a long lead, as the land is comparatively of little value. The earth is dug with a *phowrah*, the common native tool, which is at once a spade and a hoe, and an excellent tool it is.

For getting water out of foundations, and for lifting water generally either for irrigation or otherwise, several ingenious contrivances are employed. For a lift not exceeding three or four feet, and where the hole or excavation is not too small, a swing-basket covered with leaves or matting is used as a bale, being swung by two men. I have often seen water lifted in this way from 12 to 16 feet, in three or four stages, by as many pairs of men, the baskets being swung together in exact time, and the quantity lifted one stage being about 1800 gallons per hour.

For higher lifts, there are three machines commonly employed — the *paecottah*, or lever bucket, used in Bengal (the counterpoise on the shorter arm of the lever being a heavy stone or a lump of clay); the *churus*, or *chursah*, common in the North-Western Provinces, a large leathern bag drawn up by bullocks; and the Persian wheel, or endless chain of buckets, also worked by bullocks, and everywhere employed in the Punjab, which has been in use in the East for at least 2000 years, and is one of the most effective and ingenious water-raising engines yet invented.

The native carts or hackeries are drawn by bullocks or buffaloes, and are exceedingly primitive vehicles. They have no springs—often no iron tires to the wheels—and, except the axle-pins, have generally no iron at all in them. But the advantage is that they can go over any rutted track that does duty for a road, and that if they do get broken, the nearest carpenter can repair them. That, of course, is the real difficulty in introducing English improvements into India, that if your fine carts or pumps get broken, who is to repair them?*

I now come to a very interesting subject to the Indian engineer, which involves several specialities, and applies more or less to all the constructions we shall consider afterwards in detail—I mean the subject of *Foundations*. Of course the general principles of constructing foundations are the same everywhere, that is, you must secure, if possible, a firm and unyielding substratum. If you cannot find this naturally supplied, you must use artificial contrivances, either by proper distribution of your weight to be supported, or by preventing any lateral spreading, or undermining from the action of water. For this purpose in England we have recourse to piling, or concrete, or iron cylinders and screw-piles. How is it in India? Well, there, as a rule, piles won't do; first, because timber

* I remember a native's answer to a remark of mine, which shows that there are always two sides to a question, even a question of improvement. His Persian wheel, near to which my tent had been pitched, kept me awake all night with its abominable creaking, and I asked him in the morning why he didn't use grease to stop the noise? He said, "I am accustomed to it, and if the creaking ceases I wake up; I know the wheel has stopped because the bullock-driver has gone to sleep, and I go out and beat him."

is scarce and dear; second, and chiefly, because it is exposed to so many causes of decay that it would quickly rot. Iron cylinders and screw-piles have been a good deal used lately; but generally in India, both these and the timber piles are superseded by the employment of Cylinders of Brick masonry, varying in diameter inside from 3 to 12 feet, and sunk either to a firm stratum below, or to such a depth as to be safe from scour,—the weight being borne in this case by the friction against the sides. A sufficient number of wells are designed to carry the superincumbent weight, whether it be a house or the pier of a bridge, and the whole series being sunk to the required level, and as close together as possible, the tops of the wells are arched over, the arches are all connected together by slabs of stone or other arches, and on this artificial platform the superstructure is raised.

To sink a well, the *néemchuk*, or well-curb, a ring of wood from 9 to 18 inches thick, is laid on the ground, the masonry built upon it about 4 feet high, and left to dry. The sand inside is then scooped out, and the well descends gradually, when another 4 feet are built up; the sand is again scooped out; and so on, until the required depth is reached. So long as there is no water met with, and the soil is sand, the work is easy enough, the only care being to see that the excavation proceeds regularly and evenly; but when the water is reached, and as it deepens, the process is a slow one. A huge sort of hoe (called a *jham*) is used, which is worked from above into the soil, and then hoisted up with its load; a

F

Persian wheel or a *churrus* being also used to keep the water down as much as possible. Sometimes blocks of wood or kunkur are met with which impede the descent of the well, and then a diver must be employed,— a man who descends without any diving apparatus, and can stay under water an alarming length of time.

Since the construction of so many great railway bridges, in which these cylinders have been sunk to a depth of 50 feet, various improvements have been introduced. The wooden curb has been replaced by one of wrought iron, with a sharp-cutting edge, into which are bolted below several rods of inch iron running through the masonry, and connected at intervals by flat iron rings, so that the whole cylinder is well bound together. A Sand Pump has also been invented, worked by a steam-hoist, to take the place of the old native *jham*. This, and another machine for the same purpose, called Fouracre's Well Excavator, you will find described in the 'Professional Papers' and the Roorkee 'Treatise.'

Sometimes, instead of a series of separate wells, the whole mass of a pier has been built in a block, with hollows at intervals, and sunk together by several parties of well-sinkers. In this way the piers of the great Solani Aqueduct, at Roorkee, were constructed and sunk 20 feet; but this work requires very experienced men, and is subject to such risks that, as a rule, separate wells are preferred.

Such work as the above can only be carried on

when the river is low, and the current very slack, and this may be a good place to speak of the general nature of such Indian Rivers. I refer chiefly to such streams as the Ganges, Jumna, Indus, and others, as there are few things of more vital interest to the Indian engineer.

They rise in the region of perpetual snow, generally from glaciers at a height of over 20,000 feet, and make their way to the plains (receiving numerous affluents on the road), as great torrents with numerous rapids, and quite unnavigable, except in particular places. On emerging into the plains, which they do very suddenly, the bed changes, first from boulders to shingle, and then to sand, and the stream cutting deep into this soft soil, becomes more and more charged with silt, especially as the lessening slope of the country checks the onward velocity. Thenceforward, the river is a sluggish stream, pursuing a tortuous course between low, flat banks, and full of shoals and quicksands, until near its mouth, where it parts into numerous channels, and forms a vast delta reaching to the sea. Owing to the vast quantity of silt brought down by these rivers, which elevates their beds and during the inundation season is freely deposited on both banks for some distance inland, it follows that these streams, in the lower portions of their course, often run on a ridge, and not at the lowest points of their valleys,—just like the Po and Mississippi, which are retained between artificial dykes.

These rivers are at their lowest in the cold season;

and in December, January, and February, their navigation is attended with the greatest difficulties. From March to June, the increasing heat of the sun melts the snow in the higher ranges, and the rivers rise rapidly. In June, July, and August, the monsoon rains increase their volume to a prodigious extent. Their banks are often inundated far inland, and the yellow turbid waters carry down more silt than ever; the increased velocity of the current, acting on the light sandy soil of the sides and bed of the channel, cuts away the banks and scours out shoals in one place, and then at the next bend a temporary check will throw down enormous quantities of this silt, and the stream shoots off in a new direction altogether. I have seen the whole of the Indus concentrated in a not very deep channel 1000 feet across; six months later I have been in a boat on the same spot, and was unable to see either bank from mid-stream. In one year I have known it cut its way inland a full mile, measured perpendicular to the thread of the stream.

It is these violent changes in velocity and direction, and the soft and treacherous character of the soil, that make the question of foundations in water a peculiarly difficult one in India. The sudden shifting of a shoal may dam up the archway of a bridge, and the increased velocity in the narrowed waterway acting on such soil, often scours out a hole 30 feet deep in a single night. Almost every rainy season in India sees failures and disasters on this score, and the most carefully considered design may be a mass of ruins in twenty-four hours. The Bengal mode of dealing with these difficulties is to carry down every pier-cylinder, whether of masonry or iron, to such a depth as either to rest in the firm soil below, or to be absolutely beyond the reach of any possible scour, but to put no flooring or curtain walls by which the stream can possibly be checked and incited to tear up the bed. The Madras system is to be content with a much smaller depth of foundation, but to provide, by means of a flooring and apron walls, a compact dam of masonry solidly bound together, front and rear and from shore to shore, which shall be proof against any action of the stream. The subject is one which has excited much discussion, but is too long to enter upon here. You will find it treated of in the works already quoted.

I have now described the most important specialities of Indian foundations, and have to ask you to follow me in the first section of what I may call special engineering constructions, *viz. Buildings,* or rather dwellings, such as houses, barracks, churches, and the like. The subject is, of course, a very extensive one, and I only propose here to draw attention to those buildings which the Indian engineer is generally called upon to construct. First, then, let us take Indian Barracks, that is, barracks for English soldiers in India, as they are not only the most important buildings that you will have to deal with, but there are many vexed questions concerning them which are still far from settled, and which apply equally to all dwellings erected for Europeans in an Indian climate.

In the earlier days of our Indian empire, the barracks constructed for the European soldiers were often built on ill-chosen sites; the rooms were low and a great deal too crowded, and drainage and conservancy scarcely thought of; the consequence was a frightful mortality, which often reached the high figure of 70 to 90 per thousand. As the important subject of sanitation received more and more attention in England, it was not likely it should be neglected in India, where a tropical climate aggravated the results of neglect of sanitary laws, and where every English soldier who died cost the Government a large sum of money to replace him.

At the same time that attention was thus roused, the conquest of the Punjab and the necessity of quartering a large number of troops in that province to defend the most exposed frontier of the empire, made the subject doubly important, while it gave a field on which to carry out the results of recent investigations. Large sums of money were accordingly spent on the barracks at Mean Meer, Sealkote, Nowshera, and other new stations. They were all single-storied buildings—each company of 100 men having a barrack to itself, consisting of six wards, each 48 by 24 feet, and 24 feet high, with double verandahs on both sides, besides rooms for the non-commissioned officers. The men were to sleep in the wards, and to dine in the inner verandahs. A reading-room was also supplied at the end of each barrack. Wash-houses and privies were arranged in separate buildings near the barracks.

These buildings were certainly a great improvement on anything hitherto erected, and were decidedly not open to the charge of being overcrowded. Unfortunately, however, the Punjab has a cold season as well as a hot, and in the winter months the men complained that it was almost impossible to keep themselves warm; while even in hot weather, the high winds and dust found their way through the numerous doors and windows, and the barracks were anything but comfortable residences.

Nor did they prove at all healthy, for cholera and intermittent fever played sad havoc among the troops in more than one season at Mean Meer. This was not, however, owing to any fault in the barracks, but to what was an error in the choice of site for the cantonment; and attention is drawn to it because the instance is instructive, and has more than once been repeated. Apparently under the idea that there was something in the very presence of vegetation which might engender malaria, a flat, bare, dreary-looking site was selected, totally devoid of all vegetation, and with water 40 to 50 feet below the surface; while, from the presence of a kunkur or stony stratum a few feet below the top, it has been found impossible to get a tree to grow to a respectable size. It may be taken as a rule that the presence of healthy vegetation shows a site favourable to human beings as well as to trees and crops, and that such sites as that of Mean Meer should be carefully avoided in future.

Several committees were subsequently appointed to consider the question of barracks, and finally an officer of experience, under the title of Inspector-General of Military Buildings, was appointed to collate the various information collected, and to prepare standard plans of barracks for the various stations at which the Government of India had determined, on strategic grounds, to maintain a permanent force of British soldiers. It should be remarked in reference to this point that erroneous impressions often prevail in the public mind. It is often asked, why are troops retained at stations which are notoriously unhealthy? Why are the British troops not all kept at the hill stations, where they enjoy a European climate all the year round? Now, the first use of having troops at all is unquestionably that they should be ready for action at the points where trouble is likely to arise; those points are the great cities where, in the East as in the West, the *mauvais sujets* of the empire are generally found congregated; and all our principal cantonments have been located with this view. It is true that the railways have done much to abridge distance in India, as elsewhere. But not to mention how easily a railroad is disarranged in time of war or revolt, the railway system must be extended at least five-fold before it could help us much on this point, while to allow revolt to rage unchecked, for even twenty-four hours, in such cities as Patna, Benares, or Lucknow, until troops could be brought by rail from the nearest hill station, would be a frantic proceeding.

But the fact is that it is very questionable whether a continued residence in the hills would be beneficial to the health of all English soldiers, while there is no doubt that the hill stations are not at all popular with the men themselves. Private Tommy Atkins is insensible to the poetry of mountain scenery; he dislikes the incessant rain from June to September; he hates being unable to take a walk without having to go up or down hill; and he misses the excitement of a large native city. There is no doubt, however, that, as *occasional* residences, the hill stations are most valuable, and the Government has shown its appreciation of this fact by the large expenditure now being incurred on the two new Himalayan stations of Chakrata and Raneekhet.

You will find plans of the most approved forms of barracks, as at present sanctioned, in Vol. II. of the Roorkee Civil Engineering Treatise. These buildings consist of a single main ward, with one verandah all round, and are to be double-storied as a rule; the upper story being used as a dormitory, the lower for day-rooms. Separate rooms are provided for the sergeants, and separate sets of quarters, of course, for the married men. In the plains, space is to be provided at the rate of $7\frac{1}{2}$ running ft. (of wall), 90 superficial ft., and 1800 cubic ft., per man, the wards being 24 ft. high. These buildings are constructed of brick or stone, as the case may be; the roofs are iron girders with brick arches between for the verandahs, and iron trusses covered with a double layer of tiles for the main wards.

Several ranges of barracks on the Government standard plans have been erected at Allahabad, Saugor, Jullundur, Peshawur, and other stations; but it cannot be said that the results have been at all commensurate with the large cost incurred. The heat and glare are greatly complained of, and in more than one instance it is believed that a return to the old temporary thatched barracks was urged. Nor have the new barracks apparently proved more healthy than the old ones; cholera stuck so closely to them at Allahabad that for some months they were absolutely deserted, and it is understood that Peshawur is no better off. Of course, it is not meant to suggest that the new barracks engendered cholera, but merely that a site unhealthy in itself has not been made less so by the erection of large and expensive buildings.

In the station where I have been quartered for eight years, the barracks were little more than temporary sheds run up just after the Mutiny, and which have not been replaced by regular buildings because it is not thought necessary to retain troops there permanently. Yet in these crowded sheds, troops have been healthier than in any station in the country.

To understand the special difficulties attending the construction of Anglo-Indian dwellings, it is necessary to revert again to the characteristics of the climate of Northern India, in which by far the largest number of our troops are quartered. The seasons of Northern India (say from Benares up to Peshawur) are the cold, the hot, and the rainy season; or, more exactly, the cold, the dry-hot, and the moist-hot seasons. In November, December, January, and February, though the sun is always powerful, the temperature of the air is pleasant during the day, and not too hot for a long day's cricketing or shooting on foot; the nights are really cold, and occasionally even frosty. A good deal of rain also falls in the Punjab generally about February, and fires are required in the house often all day long.

In March, the hot winds begin to blow; strong breezes from the west, often rising to gales, bring up clouds of dust, and the atmosphere has the temperature of a glass furnace; from 7 A.M. the wind blows often up to midnight, and the temperature of the outer air at night varies little from that of the day. I have often known it 100° at Lahore at *midnight.* This lasts till the middle of June, when thunderstorms usher in the south-west monsoon, or rainy season; rain falls in torrents, the air is saturated with moisture, and though the thermometer falls several degrees, the languid, moist heat is far more trying to the constitution than a higher temperature when the air is dry. The rains end in September; October is generally fine, the mornings and evenings are cool, and then follows the cold season already described.

Now to apply these facts to our subject.

As a mere question of comfort, it has been found that a building constructed with walls of mud or sun-dried brick, with a good verandah all round, and with a thick roof of thatch, is certainly the most comfortable. It is cooler in the summer, warmer in the winter, and a good thatched roof is perfectly water-tight. But such a roof is always exposed to danger from fire, and both roof and walls require perpetual repair in the rains; while those Indian pests, the white ants, besides snakes and other vermin, increase and multiply to an alarming extent in such a building. Stone is scarce and expensive, except in a few localities, so brick is the material usually employed for permanent buildings, and the thatched roofs are replaced by tiles or concrete. Such buildings, however, become insufferably hot during the day, and numerous doors and windows are provided to allow a sufficient volume of cool air to enter during the cooler hours of the night; they give rise to one of our chief difficulties, for, during the whole of the hot weather, these openings must be kept closed for ten or twelve hours at least, to exclude the heated external air, and if so closed, some special means of ventilation ought certainly to be provided.

It is not, however, easy to effect this, for the simple reason that the air inside the building is cooler, and therefore heavier, than that outside, and thus will not pass out easily by apertures above, which also generally admit quite as much dust as fresh air.

As these numerous doors and windows evidently give access to a great deal of heat, even when closed, it seems desirable to enquire whether fewer openings would not be a better arrangement, and there can be little doubt that this would be the case, if a proper system of ventilation were at the same time contrived by blowers, or otherwise, as will be hereafter noticed. It is also probable that thicker walls of hollow bricks, or even double walls, would be a great improvement over the present thin walls of solid brick. But brick-making machines are still in their infancy in India. The tiled roofs are also doubtless a great cause of the heat of these barracks. And as thatch is out of the question, it seems very desirable to find some non-conducting, non-inflammable substance, which should be interposed between the tiles and joists.

It may be noted that, amongst the natives, underground apartments are not uncommon, to escape the great heat of the summer. In the fort at Lahore, an extensive system of such *tykhanas* (as they were called) was found on our capture of the place, and they were certainly much cooler than the rooms above ground; but the difficulties of providing proper ventilation and light would be great, and they would probably not be healthy for Europeans.

The subject of what is the best description of building, both as to design and materials, for Europeans in such a climate as India, 'is indeed still an open one, and admits of great discussion. Neither our barracks, churches, nor private houses are as yet satisfactory, and as to their architectural appearance, the less said about that the better. The Gothic churches generally built in our Indian stations are insufferably hot. In that at Mean Meer, it was not uncommon to hold the service, even in the early morning, outside the church and under its own shadow! Most of the private houses in the various cantonments are simply hideous, and even the best of our public buildings, law courts, town halls, &c., are anything but adapted to the climate. Imitations of Classical, Italian, and Gothic architecture are plentiful everywhere, but few attempts have been made to adapt any of the features of Oriental architecture to our Western requirements. Yet who can gaze on the beautiful domes and minarets of the Taj Mehal or the Jumma Musjid, or the graceful arches and bold cornices of the Motee Musjid, without admiration and envy? It is much to be desired that the whole subject should be taken up and carefully studied; but so long as we have so little originality even in England, we must have patience in India.

To return to our barracks. During the hot season, the air inside is rendered more bearable by the employment of punkahs or pendant fans, which, being swung from the roof and pulled to and fro by manual labour, afford a very grateful relief, though, of course, they do not really lower the temperature of the air. Besides this, during the dry heat, *tatties*, or screens of grass, are hung in some of the doorways to windward, and, being kept saturated with water, fresh air is thus introduced, which is often, during a strong breeze, from 10° to 20° below the external temperature. But this contrivance is of course useless during the rainy season, when the external air is already saturated, and when often there is not a breath of wind stirring.

About three years ago a committee, of which I was a member, was appointed at Roorkee to consider the question of ventilating and cooling barracks and other public buildings in India. It was pointed out that the present arrangement of punkahs and tatties cost Government annually a very large sum; that the results were not believed to be at all commensurate with the cost, and that it was desirable to see what improvements could be introduced.

Many interesting experiments were made, both as to the best modes of hanging and swinging punkahs, and as to the possibility of dispensing with them altogether by using blowers, similar to those employed for ventilating mines, which are to some extent already used in India under the name of thermantidotes. It was found that a system of blowers, with tubes for conducting the air into the barrack-rooms, would certainly be useful for the purpose of ventilation, but that they would not enable punkahs to be dispensed with, for they could not produce the same sensation of coolness as a punkah, except by keeping the whole body of air in the room in movement at an extremely rapid rate by an extravagant expenditure of power, or by throwing cooled air in, in sufficiently large masses, as in the case of a tattie with a strong breeze blowing. Now, although it was easy to produce the breeze artificially during the dry heat, it was of no use during the rainy season, unless some artificial means could be found of cooling the air otherwise than by evaporation. And there appears no means yet known to science of effecting this in such quantities as are here required, and at such a cost as will not be ruinous. It is true that many schemes have been proposed, and some tried successfully, on a small scale, such as the use of ice, ether, ammonia, &c.; also condensing machines, such as Dr. Arnott's; but it is believed that the problem has never been solved on a large scale yet.

Of course, the utility of blowing in fresh air for ventilating purposes is not questioned. It has been carried out in several public buildings lately erected in India; but to blow it in even during the three hot dry months, in such quantities as to cool, as well as ventilate, the whole interior of a large barrack, is expensive work; for a whole set of barracks, comprising from fifteen to thirty different buildings, separated from each other by considerable intervals, several steam engines and an enormous quantity of tubing would be required; and the cost of such things in India is almost ruinous.

It seems probable, therefore, that the old system of punkahs will be found, for a long time to come, the most efficient and economical; but their form, construction, and mode of pulling have been the subjects of much discussion. As to the form, the committee above mentioned recommended, after a variety of experiments, that the punkah should be a rectangular framework of wood and canvas, 12 to 18 inches wide, with a heavy fringe 18 to 24 inches deep; that a system of punkahs should be rigidly connected so as to swing evenly without jerking; that the best effect was produced when they moved through an arc of 5 feet, with a velocity of $2\frac{1}{2}$ feet per second; that no machine had been brought to their notice equally effective and economical with a man's arm, but that a heavy pendulum, if properly connected with a series of punkahs, might be useful to ensure regularity of swing.

The number of machines or contrivances for pulling punkahs, of which models, drawings, or sketches were forwarded to the committee, was extraordinary. A few were ingenious, but complicated and apt to get out of order; the majority were designed by men ignorant of the first principles of mechanics, who thought that a pendulum was a prime mover, or that the weights of a clock wound themselves up.

Much of what I have said above in regard to barracks applies, of course, to other buildings as well. As Government engineers, you may have, besides barracks and their subsidiary buildings, to build or repair courthouses (commonly called *kucherries*), rest-houses of various kinds on roads or canals for the accommodation of the establishment, possibly a museum or a college, or a lieutenant-governor's palace; and chapels and churches. In all these buildings the executive engineer is often called upon to prepare the original design, and if he has any architectural skill, may have an opportunity of distinguishing himself. As I have already hinted, I cannot say that we have succeeded in our attempts at Anglo-Indian architecture. The leading idea a few years back was, apparently, that a barrack was the unit, or standard, or germ of all architectural designs; that a barrack was necessarily a rectangular parallelogram with four walls and a roof; that the addition of a stuccoed Grecian portico, with a

G

cross and green venetians to the windows, produced an excellent Roman Catholic chapel, and that if you wanted a Protestant church, you had only to omit the cross and add a tall, square tower. However, we have grown beyond that now, and produce excellent copies of Gothic churches, which would be very nice if they were not so hot; and in some of the Bombay public buildings, there has been a considerable amount of originality in the adaptation of the Gothic to Indian requirements. There is, however, still plenty of room for any young candidate for architectural honours to develop original talent in this direction. I would only say to him, for goodness' sake try to be original, and don't be content with slavish copies of buildings erected in Europe four or five hundred years ago for a totally different climate. If you *cannot* be original (and doubtless the genius of originality is given but to few), then study the noble specimens of Eastern architecture that are still left to us, both in India and other eastern countries, and strive to comprehend the meaning and intent with which that style was designed; you may then catch something of the spirit of those great builders, and produce something at least suitable to the climate and the country, and creditable to the taste, which does not form a grotesque excrescence out of harmony with everything around it.

Private houses are almost invariably one-storied in Upper India; the rooms are at least 20 feet high, with numerous doors opposite each other, and small windows above for ventilation purposes. Each bedroom has always a bath-room attached to it. A verandah runs all round the house. The roof is either flat and covered with stucco supported on beams and joists; or it is pitched and covered with thatch, which is much the cooler arrangement. The floors are of polished lime, called chunam; the doors are double; the walls are white or colour-washed—never papered, as it is difficult to get the paper put on properly, and it is apt to peel off in the rains. But there is no reason that it should do so; and I strongly recommend you to get a practical lesson in paper-hanging and wall-colouring and painting before you go out, for you will find the knowledge exceedingly useful; it is simply impossible to get that kind of work done out of the Presidency towns.

Such buildings as racket-courts and swimming baths you may also have to erect occasionally; but I don't know that there is anything special to say about them, except that the former are always open, with white walls and black balls, and that the front wall should look to the west, so that you may be screened from the sun in the evening. Of course you can have a morning court on the other side, if the station is rich enough to afford it.

Every executive engineer is supposed to be consulting engineer and architect to the public generally, and the civil authorities of the district in particular, and if the magistrate is an energetic man (as he generally is), you may have to prepare designs for market-places, *serais* (or resting-places for native travellers), municipal offices, clock towers, and the like; and if you are of a mechanical turn, you may be sure that you will be consulted about the machinery employed in the jail manufactures, where you will find some mechanical appliances that will considerably astonish you.

The Water Supply of a range of barracks is, as a rule, derived from wells, which indeed is the only practicable course in the very flat plains of Upper India. The water is generally good and wholesome, and the supply sufficient. This is, indeed, the general source even for all Indian cities. In only a few have proper waterworks been as yet constructed, by which a supply can be delivered under pressure. Even in Calcutta, the works are not yet completed.

The question of a proper water supply for native towns is daily assuming more and more importance, especially since it now seems pretty clear that there is a close connection between impure drinking water and that terrible scourge the cholera. Waterworks have been designed and carried out in the case of a few important towns like Poona, but the majority of towns cannot afford to pay for expensive conduits, filtering beds, pumps, and reservoirs, while they might be able to pay for simpler schemes. In these as in other similar cases, I would impress on you the necessity of mastering principles and thinking out of a groove —I don't mean neglecting details, but turning your attention to what is essential, and ignoring what is merely accidental.

For the privies, the dry-earth system of conservancy,

common all over the East, is universal, and, if properly carried out, is no doubt the best and healthiest. Such a system, however, it is almost impossible to work properly in a large crowded city, and there is no doubt that a regular system of sewers is greatly to be desired in such cases. That of Calcutta is only partially completed. In Bombay and Madras, complete schemes have been devised, but I believe are not yet commenced. The cost of such works is again the obstacle, as well as the prejudices of caste among the people; the want of surface fall in so many cities is also a difficulty, as great expense would have to be incurred for pumping.

There are no Gas Works except at the Presidency towns, and as coal is only found in a few places in India, other cities must wait until some one can devise a mode of manufacturing gas cheaply from vegetable oils, which seem the most practicable source of supply in India. The subject is an important one.

India abounds in rivers, streams, and watercourses of various kinds, and therefore the subject of *Bridges* is a very important one to the engineer; but before describing any of the permanent structures that have been built, it may be useful to say a word as to the temporary expedients employed on lines of road, when money is not forthcoming for the more expensive forms of construction. The first of these is a sort of Irishman's bridge—it is no bridge at all; you go through the water instead of over it, by means of Paved Causeways, which are often employed on watercourses which contain no water for the greater part of the year, and are only flooded occasionally. The banks are cut down to a gentle slope on each side, and a pavement or solid flooring of masonry or concrete is built to afford a firm roadway for vehicles. If the water is too deep or the stream too strong, the traveller must wait till the flood goes down. You will find a good account of such a pavement across the River Soane in No. 2 of the 'Professional Papers.' It is a mile long by 12 feet wide, and answers its purpose very well.

On many of the great lines of road, Boat-Bridges are in use during the low-water season, which are taken up and replaced by ferry boats during the rains. The boats used are the ordinary native boats, with a platform laid on balks and saddles; but they are often troublesome when the river is falling, owing to the necessity of removing some of the boats before they are left high and dry. Cylindrical Pontoons are preferable in this respect, and are often used. There is a pontoon bridge over the Jumna at Agra, and a very fine one at Cawnpore, over the Ganges. You will find descriptions of these in the 'Professional Papers.'

On the Hill roads, temporary bridges are very common. Some are very simple, consisting of a single suspended cable, on which is a sort of travelling cradle in which the passenger sits, and is hauled across by tackle from the shore end. Others consist of a rope for the feet and two others for the hands, the three being kept apart by triangular sticks, and if you are of an acrobatic turn of mind, they are convenient enough. More ambitious specimens have a suspended platform of bamboo 3 feet wide; and a neat Wire rope bridge of 200 feet span was built over the Jumna two years ago by some young officers who went out, with the help of the native sappers.

You will find another peculiar kind described in the 'Treatise,' and in No. 166 of the 'Professional Papers,' put up by Major Lang, R.E., on the Hindustan and Thibet road. These are *Sanghoos*, consisting of beams weighted with stones, and gradually projecting one over the other until they meet in the centre. A Rope Suspension bridge worth study, over the Chenab, is described in No. 202 of the 'Professional Papers.' I would also draw your attention to the account of the Iron Suspension bridge over the Beosi at Saugur, described in No. 30 of the 'Papers,' erected forty years ago by an infantry officer with the unskilled labour of the district, and using materials procured on the spot. It is a good specimen of the way in which an officer is often called upon in India to exercise his mother-wit, and dispense with the ordinary means and appliances.

Of more permanent bridges, Timber structures are not much used in the plains, because wood is generally very expensive, and the extreme heat and damp loosen the joints and threaten decay to the timbers. You will, however, find several described in the books so often referred to, notably one over the Barra river, near Peshawur, by Lieutenant (now Major) Browne, R.E., one of my assistants at Roorkee lately.

Masonry Bridges of brick and stone are common enough, and you may often have to construct them. You will find several described in the 'Papers,' and the descriptions of the Morhur and Markunda bridges are especially good and suggestive. The piers and abutments of such bridges are generally founded on well-cylinders in the way I have already described, the number varying from three to ten according to the weight and size. In the bridges on the Lahore and Delhi Railway, great economy of construction was attempted by making each pier of one single well-cylinder, $12\frac{1}{2}$ feet diameter inside, supporting a pair of lattice girders carrying a single line of rails. Now, as these were sunk from 40 to 50 feet below the bed, and had an additional height of some 20 feet to the girder above, the stability of such a long slender column was somewhat doubtful; and when the river scoured out its sandy bed down to the clay stratum in which the cylinder rested, the force of the stream, acting with such a leverage, threw over several of the piers;—at least, in this way I interpret the failures that occurred in these bridges last year. Had each pier been composed of two cylinders (one for each pair of girders and line of rails), with the two braced diagonally together, they would probably have stood.

I have not spoken of the calculations necessary for the Discharge of these rivers, and the Waterway of these bridges, because it is too long a subject to be treated of in a lecture. I will only say here that in no one single point is the Indian engineer so liable to make mistakes; there is nothing that requires so much study, and failures on account of insufficient waterway are of every-day occurrence in that country. And even when your bridge is built, and you have good reason for believing the waterway to be ample, so treacherous are these streams, that their course may change in a single night, and your abutments may be taken in reverse. In the bridge over the Sutlej, thirty-eight spans of 110 feet each were provided for the stream, and yet, when I saw it, not a drop of water was going under the bridge; the stream had shifted while the bridge was being built. Twenty other spans have since been added, the embankments have been made very massive and defended by spurs and groins at a great expense against the action of the river; yet there is no certainty that the river may not attack the embankment at some other point in the breadth of its valley (which is 5 miles across), and force its way through, leaving the bridge high and dry. By the last mail I see that one of the piers has failed, and all traffic is stopped.

Of Iron Girder Bridges there have been many fine specimens erected since the railways have been begun. Except a few small ones which have been constructed at Roorkee, they have all been made up in England, sent out in pieces, and erected on the spot. I especially commend to your notice the paper on the Tonse Bridge, in the 'Professional Papers,' as giving an excellent idea of the difficulties of this kind of work.

So much, then, for bridges. I now turn to the very important subject of *Roads*. Great progress has been made within the last twenty years in providing India with roads, but thousands of miles are still wanted before the road system can be considered anything like complete. The cost of carriage in many parts of the country is still enormous, owing to the wretched state of the unmetalled tracks, and whole districts may be starving while plenty reigns in those adjoining, owing to the same cause. However, these are questions with which the young engineer at least has seldom anything to do; his work begins when the construction of the road is decided upon. The rules regarding the laying out and construction of roads are of course the same in India as in other countries. I will only dwell on characteristic specialities. Of the construction of the earthwork I spoke in my last lecture. The Metalling employed is kunkur, already described, and various kinds of stone used in macadamizing in the ordinary way. Heavy iron or stone rollers, drawn by bullocks, are usually employed, but steam road-rollers have lately been introduced in Calcutta, Bombay, and even the wilds of the Central Provinces.

The Grand Trunk Road* runs from Calcutta to Peshawur, a distance of 1600 miles, passing through the Rajmehal hills by some very heavy works, then on to Benares, Allahabad, Cawnpore, and Delhi, thence

---

* The Grand Trunk Road having been first metalled with kunkur when Lord William Bentinck was Governor-General, that distinguished nobleman was called by some facetious engineer, William the Kunkurer.

to Umballa, and Lahore, and Jhelum, by massive embankments; thence for the last 160 miles to Peshawur it is carried by some of the heaviest and most difficult works in the world. The designer and constructor of the whole line was Lieut. (now Colonel) A. Taylor, R.E., C.B., director of the attack at the siege of Delhi.

The Hindostan and Thibet road is another fine work which you will find described in the books. It is carried right through the hills from Kalka at the foot, *viâ* Simla, up the valley of the Sutlej, and is about 300 miles long; but only the first 60 miles are open for wheeled traffic. Sometimes it is carried but little above the level of the river, then it ascends a hill-side by a series of steep zigzags, then it goes boldly through a rock by a heavy cutting or short tunnel, then through a magnificent forest of deodars, then by the side of a vertically scarped hill, 2000 feet down, in which the road is either blasted out of the solid rock or supported by iron bearers let into the rock, and carrying a timbered floor. Probably no road in the world offers such fine scenery to the traveller. The officer who did most of the work on this road was Major Lang, R.E., my *locum tenens* at Roorkee. Before leaving the subject of roads, I may draw your attention to two or three instruments used in tracing hill roads, which you will find described in the 'Treatise'—the Madras Clinometer, and De Lisle's Clinometer.

It was in 1849, I think, that the first short experimental line of *Railway* was opened in India, between Bombay and Tanna, and soon afterwards the then Governor-General, Lord Dalhousie, issued his famous memorandum laying down the principles which were to be adopted in the construction of the great imperial lines. That great statesman had been President of the Board of Trade, under Sir Robert Peel, during the time of the English railway mania, and had been an eye-witness of the reckless manner in which money had been squandered by the system, or rather no-system, which had characterized the construction of the English lines. From that at least his famous minute preserved India. It prescribed the chief lines to be laid down, the gauge to be observed, and various other points essential to prevent the public from paying for the rivalry of competing interests. The money was chiefly raised in England by the various companies, but the Government guaranteed a minimum dividend of 5 per cent. per annum to the shareholders, in return for which it was to exercise a complete control over the work, through its own consulting engineers. This divided responsibility doubtless had great disadvantages, and it has since been thought preferable for the Government to borrow the money itself, and to undertake the construction of the lines through the P. W. Dept.; but at that time Government was not in a position to undertake the work, and the money subscribed to make railways could not of course be diverted to any other purpose, as would probably have been the case in the event of war, had Government borrowed the money itself. At any rate, the main lines have been constructed, though doubtless at a higher cost than was anticipated, and their opening has been attended with incalculable advantage to India, both in a military and political, as well as a commercial, point of view.

The East Indian Railway runs from Calcutta up the valley of the Ganges, by the great cities of Burdwan and Patna, near Benares, to Cawnpore, Mirzapore, and Allahabad, and thence to Allyghur and Delhi, a distance of 1040 miles. Agra is served by a branch line from the Toondla junction. There is also a branch line, 220 miles long, from Allahabad to Jubbulpore, where it joins the Great Indian Peninsula line from Bombay, a distance of 600 miles. The Eastern Bengal line runs for 120 miles from Calcutta to Kooshtea, whence it is now being continued towards Assam. Another line runs from Bombay across India to Madras, whence a line runs down the coast to Trichinopoly. From Delhi the main line proceeds to Saharunpore, Umballa, and Loodiana, to Lahore, a distance of 330 miles, whence a line of 210 miles runs down to Moultan. There is also a branch line, 50 miles long, from Cawnpore to Lucknow, and a few others with which I need not trouble you.

The new lines now being projected, and to be constructed by the Government itself, are the line from Lahore to Peshawur, 265 miles, one from Moultan to Kotree, down the valley of the Indus, 600 miles; one from Agra, through Rajpootana, to Indore; and another from Delhi. It has been resolved to adopt a mètre gauge for these lines, and to construct them with the severest regard to economy, and the Government engineers, military and civil, have doubtless a fine field before them.

To return to the lines already opened; they comprise some of the finest engineering works in the world. On the East Indian line, are the great iron bridges over the Jumna at Allahabad, and those over the Tonse and Soane, besides very heavy embankments and severe rock cuttings in the new chord line through the Rajmehal hills. On the Lahore and Delhi line, are the heavy embankments between the Jumna and Sutlej, and the bridging of the Jumna, Sutlej, Beas, and Markunda rivers. The continuation of this line towards Peshawur by the Government engineers will involve works of great difficulty and magnitude in the rocky country between the Jhelum and Indus, and the bridging of those two rivers, besides the Ranee and the Chenab. The passage of the last river alone will task the highest engineering skill, as that of the Sutlej has already, and for the same reasons,—the exceedingly capricious character of the stream, the sandy nature of the bed and banks, and the enormous quantities of silt brought down by the waters.

On the Great Indian Peninsula line are works of at least equal magnitude, though of a different kind, foremost among which stand the great inclines of the Bhore and Thull Ghats, by which the railway ascends the steep barrier of the Western Ghats by a series of zigzags, inclines, curves, and tunnels, which render those works unequalled in the world. On the Thull Ghat, for five continuous miles there is a grade of 1 in 37, in which occur many severe curves; while the Bhore Ghat incline is 19 miles long, and in that distance has no less than thirty tunnels.

The Rails on all these lines are of the usual double-headed pattern, weighing from 66 to 90 lbs. per lineal yard, and laid generally in chairs on transverse wooden sleepers. These sleepers are either of one or other of the woods that I have already described, sometimes kyanized or burnettized; but large numbers of creosoted fir sleepers have also been imported from Europe, and a certain number from Australia. The Punjab line is almost entirely laid on Greaves's patent Pot sleepers, with which doubtless you are acquainted. The advantage of such a roadway is of course the indestructibility of its material (iron); its disadvantage, that the cast-iron sleepers are apt to get broken, and that the rigidity of the roadway is bad for the rolling-stock at high speeds. But there is no doubt that the life of a timber sleeper in India is very short, and that it is very expensive to renew them; and a good wrought-iron permanent way, that shall not be absolutely ruinous in first cost, is still a desideratum.

The Locomotives employed on the Northern lines more nearly resemble those in America than in England, as they have to burn wood instead of coal. A huge spark-catcher is hoisted over the chimney, as serious accidents at one time were common from the blazing sparks of wood out of the uncovered funnel;

and a cow-catcher is placed just in front of the smoke-box, to take up stray cows or buffaloes, who *will* get on to the line in spite of the fencing. The engine-drivers are all Europeans, I think, at present.

Fences are made of cactus or other hedges, or of the usual wire line, or often simply of a mud wall and ditch.

There is nothing particular to say about the Stations, which are similar to those in England, with some adaptation to the requirements of the climate.

The Passenger Carriages are a great deal too much like those in England, and are generally very hot and uncomfortable. The best have double roofs, with cane backs and seats; but none have punkahs or tatties; and it is only lately, and at the instance of a very stern reminder from Government, that any serious attempt has been made to render passenger carriages for European travellers more comfortable in the hot weather. On the Great Indian Peninsula line, however, there are very comfortable saloon carriages, with movable berths, and bath-room and W. C. attached. Considering that from Bombay to Calcutta or Lahore is a journey of some seventy-five hours, it is evident that these things are serious matters in India.

That the results of these railways have been of the greatest benefit to India in both a political and military sense, it would be impossible to deny. Their educational and social value, too, to the people at large have been great, and indirectly they have doubtless added much to the wealth of the country. But only one line,

I think, pays more than the guaranteed 5 per cent., so that for all the others, the Government pays a heavy charge annually for the advantages offered by the railways, and this charge has of course to be met by taxation. The traffic, too, is increasing very slowly, and it seems that much is still wanting to assist its development to a much larger extent than heretofore. One reason for this is, doubtless, that a great portion of the goods traffic consists of articles, such as grain, of large bulk and small intrinsic value, which will not bear high transit charges. Another cause is that the natives, though travelling far more than was anticipated, have not yet learned to travel to the same extent as Europeans.

One mode of increasing the inducement would, I suggest, be to make the railway fares more uniform, and the tickets more easily procurable than at present; and this mode, indeed, would apply to English, as well as to Indian, lines. There seems no more reason that you should have to buy your ticket for a railway journey five minutes before starting, and all in a scramble in front of a little pigeon-hole, than that you should be forced to purchase your postage-stamps at the General Post-office only, and just five minutes before the post goes out. I don't see why all journeys, up to a certain number of miles, should not be performed at one uniform rate of charge, and by tickets procurable at the nearest stationer's, and available for any line in the kingdom. Such a measure would be a great convenience to the traveller, would prevent him in India from being cheated,

H

as he often is, by the native railway clerks, would effect a considerable economy in the railway office establishment, and, where the lines are all under Government control, ought to be carried out with very little trouble.

But we must go on to the leading speciality of all Indian engineering, I mean, the subject of *Irrigation Works*. Of course there are other countries in which artificial irrigation is extensively developed,—more indeed than in India,—notably, for instance, in the provinces of Piedmont and Lombardy, in the kingdom of Italy, where they are in advance of us both in the economical distribution of water, and in all legislative questions affecting the administration and rights of, and property in water. But in no other country but India have works been undertaken on so gigantic a scale; and it is in the dealing with such vast bodies of water, and their carriage over such great distances and in the face of so many impediments, that the specialities of Indian irrigation works really consist.

Now, in order to give you a clear idea of the subject, I must begin by premising a few remarks of an agricultural and financial character. India is almost entirely an agricultural country. It is true she has great mineral resources, but these are as yet undeveloped. She has had valuable manufactures, and there is no reason she should not have them again; but for many ages past and for as many to come, she has been and will be almost entirely agricultural.

Again, half of the whole Government revenue is

derived from the land. It is called a land *tax*, but is really a land *rent;* that is, rent paid by the occupier or cultivator, as the case may be, to the State as the great landlord. This has been the case all over the East from time immemorial; the only ownership in the land belongs to the Government, although the occupier or cultivator often has rights of occupancy which are almost as inalienable as the right of the State.

This being the case, it is evident that the Government is as much interested as the people in the productiveness of the harvest; for if the harvests fail, the people not only starve, but cannot pay the land rent, and the Government has to feed thousands of hungry people with an empty exchequer. Owing to the absence of roads and railways in years gone by, any failure of the rains in a province produced the most appalling distress, and the people died by hundreds and thousands.

Hence, so long back as 300 years ago, the Mahomedan emperors, commonly called the Great Moguls, took a great interest in artificial irrigation, and under the direction of a Mahomedan "Royal Engineer," one Aliverdi Khan, several canals were opened out. Some of these, after falling into disuse from various causes, have been re-opened and improved under the British Government, and are now known as the East and West Jumna Canals, the one being 110, the other 500 miles, long. Then the Government commenced a great work of its own, the Ganges Canal, of which the main channels alone are 700 miles long, and the first-class distributaries 3000 miles more. Since then, the Bari Doab Canal has been completed; the Sirhind, Soane, and Sardah canals are in progress, and others are projected.

It will perhaps give you a better idea of this kind of works, which are all similar in character, if I briefly describe the largest of all, and the one with which I am best acquainted,—the Ganges Canal. This canal takes out of the River Ganges at Hurdwar, where the river leaves the hills and enters the plains, by an artificial channel 200 feet wide and 20 feet in depth, the water having a depth of 10 feet, and the fall of the bed being 2 feet per mile. This gives a volume of water of nearly 7000 cubic feet per second,—rather more than the Thames at Richmond, I think. The canal, flowing onwards, crosses three great hill torrents, two at a lower level, and one by a level crossing, and finally reaches the valley of the Solani, which it passes by an aqueduct three miles long, revetted with masonry throughout, the Solani river itself being passed by a bridge, which gives it 750 feet of waterway. After crossing this valley, 20 miles from its head, the canal arrives on level and comparatively easy ground, and pursues its way along the watershed of the country between the Ganges and Jumna rivers for another 30 miles, broken only by sundry falls and locks, and spanned by several bridges. At 50 miles from its head, it divides into two branches, of which one runs down to Futtehgurh, the main line continuing to Cawnpore, and throwing off other branches to Bolundshuhr and Etawah. From these main lines, the principal distributaries run on both sides, nearly parallel to the channel, and throwing off in their turn other minor lines,—until the whole country is covered with a network of irrigating channels, conveying the water to every field.

Thus you will see that this and other irrigating canals differ from an ordinary navigation canal, such as you have in England, in two very important particulars. First, the irrigating canal has a running stream of water which is gradually expended as we follow its course, and so its channel diminishes in size; Secondly, it is dug on the highest line or watershed of the country, so as to secure a command of level for irrigating the whole area.

Now, as to the use to which this water is applied; you will understand that there are two harvests in Upper India: the *rubbee*, or spring harvest, which is reaped in March or April, and consists mainly of wheat; and the *khurreef*, or autumn harvest, which is gathered in September, October, or November, and consists of rice, sugar, Indian corn, and various tropical products. All these require steady and constant irrigation, and they get it from the spring and autumnal rains, or from wells, during the average normal seasons; but if the rains fail and the wells run dry, they are dependent on the canals, and even when the season is regular, the supplementary irrigation from the canal greatly increases the yield of the crop.

The Soil of Upper India is generally a light friable clay, excellent for wheat, but absorbing a great deal of water. In Bengal, they have a rich mud extending to a great depth, and of such fertility that three crops are often grown on it in a year. In the Central Provinces, there is the black cotton soil, which is, I believe, disintegrated trap or granite. Across the Indus, again, we have a hard stiff clay, which, though not good for wheat, is valuable for many crops when well watered. I notice this question of the soils here, because it is an important point for the engineer when projecting a new canal, for all these soils require different quantities of water.

In designing a canal such as has been described, we must fix the head at a point high up on the course of the parent river, first to get a command of level, so that our canal may run on high ground; and next because, in the higher portion of its course, the river water is free from silt, and the river bed is more stable and less liable to the caprice of the stream, which, lower down, might abandon our canal mouth.

Having fixed on the point of departure, then comes the question, how large is our channel to be? That depends, first, on the quantity of water we can get out of the river when at its lowest; secondly, on the slope or fall we give to the bed of the canal. The minimum quantity of water is, of course, determined by observation, and this has hitherto practically fixed the capacity of canals in Northern India, because the spring crop grown in the dry season is so much the more valuable of the two; but in later projects, an additional capacity of channel has been allowed, so that an extra supply of water may be admitted in the autumn, when the river has plenty to spare.

As to the fall of the bed, or the Velocity to be allowed to the artificial waterway, that is a question depending chiefly on the nature of the soil. Several of the Indian Canals have been designed with too high velocities, and the consequence has been a cutting and scouring of the banks and beds, which have seriously endangered the various masonry works in their course. On the other hand, if you give too little velocity, of course you have to go to a greater expense by making a larger channel to carry the same quantity of water, and you won't have velocity enough to carry forward the silt held in suspension, or to prevent the growth of weeds, a serious evil in the tropics. On the whole, it seems now generally admitted that a velocity of $2\frac{1}{2}$ feet per second is about what should be aimed at, though a compromise of conflicting interests, in this as in other cases, has often to be made.

When, owing to the slope of the bed being less than that of the country, the slope, if continued, would bring the canal bed too high, and embankments would have to be made, it is evident that the bed must be lowered by an artificial Fall or Rapid, made of such material and shape that the mass of water may flow over it without injury, and take up a new level below. The best forms of falls and rapids have been the subject of great discussion amongst Indian canal engineers. The Ogee fall in use on the Ganges Canal is now superseded by the Vertical fall of Colonel Dyas,—the principle of which is that the shock of the falling water is received into a cistern sunk below the bed, so as to act as an elastic cushion. A grating of wooden bars, inclined at an angle, and fixed along the crest, is also generally added. These bars act like the teeth of a comb, and, by dividing the water into filaments, greatly reduce its force.

If you want to navigate your canal, Locks must be provided as well as falls; but the difficulties of navigation up stream are very great, and the traffic on all these canals, with a current of some three miles an hour, is in a very undeveloped state. It is doubtful whether it is not impossible to combine satisfactorily the requirements of both irrigation and navigation in the same channel.

At the heads of your branches or main distributaries, Regulating gates will be required with sluices, by which the supply of water may be divided, increased, or diminished, as the case may be. A similar work will be required at the head of the canal where it leaves the river. The tail generally ends in a fall or rapid, by which the water is discharged into a river or nullah with a scour to clear away the silt.

You will find details of all these works in the Roorkee C. E. 'Treatise.' Besides these, Bridges of communication will be wanted for the roads crossing the canal; Rest-houses (*chokees*) for the use of the establishment; Escape heads for discharging surplus water; Inlets for the reception of cross-drainage; and other works.

One chief object of carrying the canal at a high

level is that it should interfere as little as possible with the natural Drainage lines of the country; but it is occasionally indispensable that it should cross some of them, and it may cross them under three cases:—
1. When the bed of the canal is higher than the bed of the drainage line. 2. When it is lower. 3. When the two beds are on the same level.

In the first case, the canal is carried over by an Aqueduct, which may be of masonry or iron, according to the size of the canal, and which must allow sufficient waterway underneath for the passage of the drainage line when in flood.

In the second case, the drainage line is carried over the canal by a Super-passage, which is in effect an aqueduct, only that here you have to provide waterway above the canal for the river in flood. The canal, if necessary, may be carried below it by an artificial fall of the usual construction.

The third case rarely occurs in practice, and never when it can possibly be avoided—there are the same objections to it as in the case of a Level crossing on a railway. When it *has* to be done, the method of doing it may be understood by an explanation of the Rutmoo level-crossing on the Ganges Canal at Dhunowrie, six miles above Roorkee. A regulating bridge, with sluices, is fixed across the Canal, just below the junction. A series of flood-gates hung between piers is fixed across the Rutmoo, on the down-stream side of the canal; these gates fall outwards by a hinge at the bottom, and are held upright by a catch, which can be easily knocked out, and when thus held up, they retain the canal at its normal level. When the Rutmoo is in flood, the Canal sluice-gates are lowered to prevent the flood water rushing down the canal, and choking it with silt or débris. At the same time, the Rutmoo flood-gates are lowered by knocking out the catches, and the whole flood water pours across the canal, and flows down its own channel. When the flood is over, the sluices are raised, and the flood-gates also one by one, the pressure of the water being taken off by dropping planks down grooves in front, provided for that purpose. Such a work of course requires the presence of an establishment to manage it, and is very liable to be damaged by carelessness on the sudden occurrence of a flood.

Well—the canal having been made and filled with water, how are we to get the water into the fields? First, it flows down the main distributaries, and then into secondary channels, which are laid out at proper slopes, and of a size varying according to the amount of water and the area to be watered. From these it is passed by pipes of wood, iron, or earthenware, under the banks on to the fields, which are divided into squares (*kyarees*), into which the water is passed *seriatim*.

As there is generally less water than is required, village watercourses are all filled in their turns; thus one set will be filled on three days of the week, and another set on three other days, and so on. The water rate is levied by the canal officer according to the area irrigated, and the kind of crop taking it, as some crop want more water than others, and are also better able to pay for it. Such a system is obviously open to many objections, for it entails endless labour in measurement, perpetual watchfulness, and much dispute and interference, besides great waste. Some system has therefore long been sought by which the water should be measured out at the head of the distributary channel or village watercourse, and charged off like beer or wine. Hitherto, however, all attempts to do this in India have failed; chiefly, because we cannot find any practical machine that will not get out of order and cannot be tampered with, by which a uniform discharge may be secured in a given time under a varying head of pressure. Could this be done, it is obvious that if a head or opening of a given capacity were left open for a certain time, we should know exactly how many cubic feet of water had been sent to a particular village. Lieut. Carroll, Royal Engineers, a very clever and promising officer, invented a module (as it is called), which you will find described in the 'Treatise,' and which has been the nearest approach to a machine of the above kind.

By comparison of the discharge of his canal at different points and measurement of his irrigated areas, it is obvious that the canal officer can collect data for determining what is styled "the irrigating duty per cubic foot." This varies much with the soil, the rate of evaporation and other data, but on the best-managed canals, will amount to the irrigation of 300 acres for each cubic foot per second of water discharged at the main head. This is the rate on the Eastern Jumna

Canals; on others it is not nearly so high, and in new projects about 200 are usually assumed. There are various other statistical data for canals of great value, which you will find treated of in the 'Professional Papers' and the Roorkee 'Treatise.'

In Southern India, the canals are of less elaborate construction, owing to the more permanent character of the rivers in the lower portions of their course, and to the abundance of excellent stone found there, while in Northern India we have to be content with brick. The chief object of engineering skill in a Madras system of canals is the *Anicut* or Weir thrown across the river at the point of departure, by which a head of water is secured for the supply of the canals. Many of these weirs are elaborate and costly affairs—that over the Godavery is a mile and a half long, and the weir over the Soane, now under construction, is little less. I show you on the plan the one over the Kistna, which is a good example of its kind. You will see that the body is composed of rubble stone, defended by ashlar from the action of the water, and resting partly on shallow wells, and partly on the bed of the river itself. This shallow character of foundations on sandy beds, to which I have already alluded in a former lecture, is certainly a peculiarity of Southern Indian engineering; but you will see the long slope or apron of stone in rear of the weir, where the action of the water is most severely felt after being checked in its onward flow, and caused to whirl and scour. Very severe action takes place here for some time after the weir is built, and large quantities of loose rough stone are thrown in from time to time to fill up all gaps in the slope, until the whole work has become firm. Moreover, the sand in these rivers is of a coarser texture than that in Upper India, which, when saturated with water, becomes a mere quicksand, which would swallow up a much greater quantity of stone than the other. The weir, you will observe, is provided with sluices, by which the accumulated silt in front of the heads of the canals may be scoured through the body of the work, these sluice-heads for the supply of the canals on both banks of the river being built at the two ends of the weir. This system has generally been applied to the deltas of the Madras rivers, and the more gentle slope of the country has enabled these canals to be largely used for navigation as well, to the great advantage of the traffic; the fall of the bed is often only six inches per mile, or even less.

But besides its canals, Southern India, as well as Central and Western India, make enormous use of Tanks or reservoirs for the purpose of irrigation, the undulating and broken nature of the country as clearly indicating their use, as the flat plains of Northern India appear to demand canals. Many of these tanks have been in use for ages, and in the single province of Mysore alone there are not less than 30,000; hence, of course, they vary much in size, from those with an area of a few square yards up to those which contain several square miles, and which form artificial lakes on which you can navigate. The revenue of the country depends on the maintenance of these tanks, as in the case of canals elsewhere, and the chief duty of the Madras engineers is their inspection and repair, in concert with the civil officials of the district.

However much tanks may differ in locality and size, they may all be said to consist of the following parts: —1. A Dam or bund, natural or artificial, or both, by which the water is restrained. 2. A Stream, by which the tank is fed, unless it is fed by direct rainfall. 3. A Sluice or sluices, by which the ponded water is drawn off to be used for irrigation. 4. A Waste Weir, overfall, or *Calingula*, by which, when the tank is full, surplus water can escape safely. The Dam is generally the most important work in a tank, and the determination of its site, and the construction of it when chosen, often demand engineering skill of a high order. It is, of course, first of all necessary to take a very careful series of levels in order to determine whether the amount of water that can be ponded up by a given length and height of dam will be enough to pay for the cost of the work. Next, you must determine whether there is a sufficient area of land in the vicinity demanding water, and whether you are pretty sure of a good supply of water for the tank.

The Dam may be built of earth or stone, or of a combination of the two; and in the case where a great depth of water is ponded up, the foundations of the dam must, of course, be constructed in the most substantial and careful manner, for a failure may involve not merely the destruction of the tank, but the loss of hundreds of lives. The question of the best form of retaining wall to withstand the pressure of great depths of water, has excited a good deal of discussion, and you will find some of the best sections in the 'Papers,' as well as a good description and diagram of calculation of the Moota Dam in Bombay.

The other chief point in a tank besides the dam is the Waste Weir, which is generally built at the exit from the tank of the original channel by which it is fed. It must be made of sufficient length to discharge the maximum quantity of water which your calculations have shown you is likely to pass over it, and it must be strong enough, both in body and foundation, to resist both the pressure and the shock of this water. For this latter purpose, the principle of the vertical canal fall, already described, is often used, and the falling water is received into a cistern, by which its shock is broken.

You will find a chapter devoted to tanks in the 'Roorkee Treatise,' and several examples of the different kinds employed.

As to the irrigating duty of tank water, it is roughly calculated that a cubic yard of water is required for every square yard of land; this, however, refers to rice, which takes far more water than wheat. We have not time to pursue the subject further.

Before closing the subject of irrigation works, I may perhaps be allowed to name one or two of those engineers who have added most to our knowledge of that science. Foremost stands the name of the late Sir

Proby Cautley, F.R.S., of the late Bengal Artillery, the designer and constructor of the Ganges Canal; then of Sir A. Cotton, of the Madras Engineers, his great rival, head and founder of the Madras school of irrigation; Colonel Baird Smith, F.R.S., chief engineer at the siege of Delhi; and then the late Colonel Dyas, of the Bengal Engineers, one of the most scientific officers in India.

Though not exactly under the head of irrigation works, the question of *River Works*, or river improvements, is not very far removed from it, and though we have but little time to spare for it, it would not be right to exclude such a subject altogether from your notice, for many of you will probably have a good deal to say to it sooner or later.

An engineer is generally called upon to devise works for the improvement of a river, either to increase its navigable facilities or to prevent its waters from inundating the country. With regard to the first object, the work in India generally consists in removing obstacles, such as shoals, kunkur banks, sunken trees or boats, and I cannot do better than draw your attention to the admirable paper on the Gogra River Works, sent to me by the late Lieutenant Carroll, R.E., whom I have already mentioned. The systematic improvement of rivers by the lock and dam method, as in the Thames and other small streams, is hardly applicable to such rivers as the Ganges and Indus, except at a cost which puts it altogether out of the question. It is greatly to be regretted that it is so, for the navigable capabilities of Indian rivers are far inferior to what we should suppose, judging from their length and size; much inferior to those of America, for example, though the Mississippi has much the same general characteristics as the Ganges.

During the rainy season, Inundations in the lower part of the course of many of these Indian rivers often extend far inland; the delta of the Ganges is, in fact, a vast sheet of water at that time of year, and in the dry season it is an enormous jungle intersected with watercourses, and inhabited only by tigers, snakes, and other wild animals, with the exception of a few wretched fishermen or woodcutters.

Higher up the country, again, rivers, such as the Damooda, are restrained within artificial embankments, which, unfortunately, have not been systematically planned, but have been erected from time to time by the different villages threatened with inundation along its course. There are often many miles of such embankment (or levees, as they are termed in America,) under a young engineer's charge, and very anxious work it is, as he has to guard a long line from the attack of an insidious foe, often with a very imperfect garrison. Such embankments as these are always constructed of earth, and when consolidated by time and covered with grass, a very thin earthen bank will successfully resist until it is absolutely overtopped, while new earthwork of twice and three times the sectional area will be soaked through and breached readily. Another source of danger arises from rats and other vermin, which perforate your embankments in hundreds of places.

Rivers are also trained, either to prevent inundations or to improve their navigation, by the use of Spurs or groins, which are much employed in India. They are generally run out from the shore at an angle of 45° with the current, so as to deflect it towards the opposite side when needful. By a series of such spurs, a long line of bank may be protected, and a corresponding series on the opposite side may often be necessary to give the stream a set in the right direction. Such spurs may be built of stone where there is any, but in India are more often constructed of a double row of piling, filled in with fascines; the nose of the spur is often revetted with sand-bags. Sometimes ropes are anchored, and trees or brushwood tied along the rope, by which the surface current is acted upon sufficiently for the purpose. In the Markunda River, near Umballa, well-cylinders were sunk down at a great expense, to act as anchors, and were connected by stout iron chains or wire cables, to which trees and branches were attached. The advantage of spurs founded on the bottom over floating spurs or breakwaters is chiefly in the great deposits of silt which occur both above and below the latter, and which soon form an adequate protection to them from the stream.

For all this kind of work it is impossible to lay down any fixed rules; experience and local knowledge are our chief guides, though it may be said generally that rivers are like women — more easily led than driven!

I have now run over — very hastily and imperfectly, it is true — the chief branches of civil engineering which those going to India may be called upon to engage in. There are several other important branches, such as Lighthouses, Harbour works, Gas, Water, and Sewage works, on which I have not touched, both because you will rarely have anything to do with them, and because there is very little that is *special* about them so far as India is concerned. But before I conclude, I wish to give you a brief account of the Indian Survey, partly because the results of that survey are of great importance to the engineer, and because many of our own officers are employed in it.

The *Indian Survey Department* is divided into the three branches of the Great Trigonometrical, the Topographical, and the Revenue, Survey branches. To understand something of the work done by the first, it is necessary to sketch briefly the history of its operations. It was in the year 1800 that Major Lambton commenced to run a line of triangles across India from Madras to the west coast, with a view simply to a local survey. In the course of this work it was discovered that there was an error of no less than 40 miles in the assumed breadth of the peninsula at that point. The original Base Line was measured near Madras, and a base of verification at Bangalore, and from this latter base a line of triangles was subsequently carried down south to Cape Comorin, and up north to Dera, 40 miles from Roorkee. This line, carried along the centre of the continent, formed the Great Indian Arc, and is the largest arc of longitude ever yet measured on the earth's surface. Another great line of triangles has since been carried from Kurrachee, in the west, to Calcutta, in the east, bases being measured at each end. From Kurrachee, again, the Indus series of triangles extends to Attock, where another base was measured, and from Attock another chain comes down the Punjab back to Dera. There are several other series of triangles with which I need not trouble you.

In the earlier days of the survey, the instruments used, especially those for base measurement, were necessarily imperfect, but Colonel Colby's compensation apparatus, used in the Irish Ordnance survey, was afterwards sent out to India, and two 36-inch theodolites, made by Troughton and Simms, which, I believe, are the very best instruments ever turned out. Subsequently, several 24-inch theodolites on the Everest pattern were sent out, besides a first-rate Zenith Sector, and many other instruments; and all the old work has been re-measured, while fresh series of triangles are yearly being added. The published results of the work show that it is singularly accurate, and may bear comparison, in this respect, with our own or the Russian survey, while in the magnitude of its operations it surpasses both. I should mention that for the principal triangulation stations in the plains, high towers of masonry are built, from which the angles are taken. The stations on the peaks of hills are marked by small pillars of stone.

Besides the ordinary triangulation, and the determination of the Latitudes and Longitudes of the chief stations by careful Astronomical observations, a series of very valuable Pendulum observations has been lately executed for the Royal Society at intervals along the Great Arc, the object being to determine the effects of local attraction on the plumb-line caused by the great mountain chain of the Himalayahs. It was while engaged in these that Major Basevi, Royal Engineers, died alone in the solitudes of Thibet.

In addition to the above, a very extensive series of Spirit-Levelling operations has been undertaken within the last few years by the Survey Department, the object of which is to have an independent mode of determining the height of the principal stations in the interior above the mean sea-level at Kurrachee, and to establish a series of Bench Marks all over the country, by which the levels taken for any engineering project may be referred to the Kurrachee datum.

Major Lambton was succeeded by the late Sir George Everest and Sir Andrew Scott Waugh, F.R.S., now retired, both of the late Bengal Engineers. The present Superintendent of the Great Trigonometrical Survey is Colonel J. T. Walker, F.R.S., late of the Bombay Engineers, who some years ago executed an admirable military survey of the Trans-Indus frontier under circumstances of great difficulty and peril. His second in command is Major Montgomerie, F.R.S., late Bengal Engineers, who received a gold medal from the Royal Geographical Society for his admirable survey of the kingdom of Cashmere, one of the completest and most difficult surveys ever executed. Among the other subordinates is Lieutenant Herschel, F.R.S., late Bengal Engineers, a son of the great astronomer, and already well known to the scientific world for his spectroscopic observations on the last two total eclipses of the sun.

The Topographical Survey is confined to those large tracts of country which, from their hilly and jungly configuration, would not pay to be surveyed in the elaborate manner employed in the plains. The instruments used for the triangulation are 7-inch Everest theodolites, the stations being always connected with the most convenient points of the Great Trigonometrical Survey. The details of the country are filled in by the Plane Table, which is similar to the sketch block, and though not so portable, is more convenient and perhaps admits of greater accuracy. Many very beautiful maps have been turned out by this Department, the head of which for some years was Colonel Robinson, late Bengal Engineers, one of the most accomplished military draughtsmen in the service, now Director-General of the Indian Telegraph Department.

The Revenue Survey is, as its name denotes, the survey of land for the purpose of revenue assessment. It takes up the chief points determined by the Great Trigonometrical Survey, runs secondary lines of triangles where necessary, and fills in details by the Gale

Traverse system, which is specially adapted to a flat, open country. The Plane Table supplements the Traverse work, and the result is a series of maps which, for minute detail, are scarcely equalled by the 25-inch English Ordnance maps, for every field is determined and even the nature of the soil and the crop laid down.

The Revenue Survey is presided over by the Surveyor-General of India, Colonel Thuillier, F.R.S., late Bengal Artillery, who has for many years devoted himself specially to this work, and to whom India is indebted for many improvements.

And now, Gentlemen, I have finished my lectures, and would fain hope that they have given you half as much pleasure in listening to them as I have had in preparing them. The field was so vast that I could easily have extended their length and their number; but my object has been rather to awaken your interest and stimulate your curiosity than to supersede your reading.

www.ingramcontent.com/pod-product-compliance
Lightning Source LLC
Chambersburg PA
CBHW021941160426
43195CB00011B/1176